FRAILNESS AMONG ELDERLY PEOPLE
(A Psychological Study)

FRAILNESS AMONG ELDERLY PEOPLE (A Psychological Study)

By
Roshan Lal Zinta
Priyanka Thakur

Neha Publishers & Distributors
4832/24, Prahlad Lane, S-207 Ansari Raod, Daryaganj, New Delhi-110002

Published by
Neha Publishers & Distributors
4832/24, Prahlad Lane, S-207 Ansari Raod, Daryaganj,
New Delhi-110002
Tel.: 011-43570976, 23278261
Email-nehapubdistributors@gmail.com

First Edition : 2013

ISBN : 978-93-80318- 27-1

Price : 350/-

Laser Typesetting by : Vinod Bhardwaj

Printed at: Vikas Computers, Delhi

PRINTED IN INDIA

Published by Neha Publishers & Distributors, New Delhi-110002

PREFACE

The concept and history of frailty date back to the 1970s with the endeavor of Monsignor Charles Fahey and the Federal Council on Aging in the United States who were credited with introducing the "frail elderly" to describe a particular segment of the population. It goes along with ageing, as inevitable as white hair and wrinkles, therefore depicts the price to be paid for living to an advanced age. Roughly 1,50,000 people die each day across the globe and two thirds of them die with age-related causes and 90% of them belongs to industrialized nations. In simple words the *frailty in its relation to adverse outcomes include dependency, disability, institutionalization, falls, injuries, acute illness, hospitalization, slow or blocked recovery and mortality.* The "cycle of frailty" involves sarcopaenia, neuroendocrine dysregulation and immune dysfunction being the "physiological triad" to the frailty. Hormesis might result from either direct stimulation or over-compensatory response to noxious stimuli. Although a variety of physiologic underpinnings have been proposed, most researchers have pointed out that the frailty results from a multi-system reduction in reserve capacity to the extent that a number of physiological systems are close to the threshold of symptomatic clinical failure. So, the frailty most often in literature represent the accumulation of deficits or as a distinct clinical syndrome involving specific co- morbid physical disorders and functional impairment or disability, whereas the psychological aspects on it has received little attention. *The person with psychological frailness shows allostatic load that involves "wear and tear" or "use it and lose it" that occurs in an organism over time to maintain a steady state.*

The term was selected to focus attention on a group of elderly with physical, emotional, and social impairments. It was used as an administrative device for triggering access to core support services for the elderly. The onset of frailty is associated with an identity crisis; a psychological stage. Mental health in older people has for too long been a secondary concern. The depression is reported in around a third of residents and dementia in two thirds in care homes. Behaviour disorder is a frequently cited to trigger for people with dementia to be admitted in a care home. The life course approach provides an attractive framework towards understanding frailty and its determinants. It integrates biological, social, clinical, cognitive, psychological and environmental factors those interact across a person's life span development.

Brocklehurst offered a "model of breakdown of frailty with assets and deficits. The assets includes health, functional capacity, positive attitude to life, caregiver availability and other resources while the deficits proposed included chronic disease, disability, dependency and caregiver burden which threaten independence. *Frail individuals are those whose deficits outweigh assets* (Brocklehurst, 1985). Lebel, in his model of frailty included predisposing factors that influence the development of age-associated degenerative changes that can cause impairments. Cognition, neuro-locomotor and energy metabolism impairments were emphasized (Lebel, Leduc, & Leclerc, 1999). There is multiple system impairment with critical changes in its reserve capacities, especially metabolic, cardiovascular, musculoskeletal, immunologic and neurological systems. The parameters of frailty deals with self-reported and performance indicators of poor hand grip strength, slow walking speed, poor levels of physical activity, unintentional weight loss and positive report of exhaustion/fatigue followed with mental vulnerabilities. Individuals with three or more of these characteristics are classified as "frail" while those with one or two are labeled as "pre-frail". Using this definition, frailty has an independent predictive validity for mortality, disability and hospitalization. The gene and chromosome defects followed with wear and tear due to lesion and damages may causes threat to the life style in old age. Further the automatic induced antibodies and defect in biological clock reduce life expectancy and wellbeing in old age. Adler once pointed out that the only standard we have for judging all of our social, economic, and political institutions and arrangements as just or unjust, as good or bad, as better or worse, derives from our conception of the good life for man on earth, and from our conviction that, given certain external conditions, it is possible for men to make good lives for themselves by their own efforts. In this book an attempt has been made to study frailty psychology so as to promote subjective wellbeing in the senior citizen. The authors have given innovative attempt to study the syndrome psychologically. In chapter one general concept has been given whereas in chapter two it has been theoretically supported. Further literature has also been framed and appropriate methodology has been narrated in a more scientific way. The study is both qualitative and quantitative in nature. The results and discussion has been presented in the forthcoming chapters.

The author thanks Mr. Sudhir Choudhary, the proprietor or Publishing Company for giving best shape to the manuscript.

CONTENTS

INTRODUCTION

I am the cygnet to this pale faint swan,
Who chants a doleful hymn to his own death,
And from the organ-pipe of frailty sings
His soul and body to their lasting rest."

—William Shakespeare, King John Act 5 Scene

Epitomizing Psychological Frailty

Like Shakespeare, most people feel that the meaning of frailty is intrinsic. They call them the frail elderly. The assumption is that frailty goes along with ageing, as inevitable as white hair and wrinkles. It is the price to be paid for living to an advanced age (Kolata, 2004). Out of roughly 1,50,000 people who die each day across the globe, about two thirds die of age-related causes. In industrialized nations, the proportion is much higher, reaching 90 % (Aubrey & de Grey, 2007). So, the frailty most often in literature represent the accumulation of deficits or as a distinct clinical syndrome involving specific co- morbid physical disorders and functional impairment, whereas the psychological aspects on it has received little attention, need microanalysis (Bandura, 1977) and humanitarian approach (Zinta, 2006, 2007, 2008, 2009, 2010) while surviving in complex technological environment. Theorists even have overlooked the need to even define "old", "older", "elderly" more rigorously. More appropriately, the aging is an integral part of human condition involves sub-cultural phenomenon therefore is innovation of early nineteenth century (Atal, 2001) that involves various socio-psychological factors. According to social theory Banton, Clifford, Erosh, Lousada and Rosenthal (1985) the social environment enters into the life of an individual and construct him socially through three economic, political and ideological practices in result can make people psychologically frail.

It is proposed that the onset of frailty is associated with an identity crisis; a psychological stage of adult development termed the "frailty identity crisis." The changing demographic profile of the world's population towards old age and evidence of people living for longer life with less time spent in ill health. "Mental health in older people has for too long been a secondary concern, acting as the 'poor relation' to other physical health problems or mental health problems in younger people need due attention. According to a research about five per cent of frail older people live in long-term care in the UK and one fifth of all deaths occur in care homes. Depression and dementia are prevalent mental health conditions in care homes. The depression is reported in around a third of residents and dementia in two thirds (Dening, & Milne, 2009). The UK has nearly 500,000 care home places in approximately 13,500 homes. Behaviour disorder is a frequently cited to trigger for people with dementia to be admitted in a care home. Hence there are approximately 80-90% of residents suffering with dementia those experience behavioural problems. Agitation is commonly used but imprecise term, comprises a range of disturbances in activity often combined with subjective distress. Aggressive behavior covers a spectrum of activity, from verbal abuse, through hitting or scratching when a resident misinterprets the purpose of care tasks.

Over the past decade, concern has been growing about society's ability to meet the mental health needs of elderly people. Certain racial and ethnic issue has become particularly salient given the rapid increase in the U.S. elderly population. In fact, between the years of 1990 and 2000, the elderly population in the United States increased 12 percent, and elderly Americans currently comprise 12.4 percent of the U.S. population (U.S. Census Bureau, 2001). Moreover, the population of racial/ethnic elderly is estimated to be increasing at a faster rate than that of whites (Ruiz, 1995). In India also the racial and socio-cultural factors are directly associated with the psychological frailness both in rural and urban areas. Despite these concerns, relatively little attention have been given to studying the mental health needs of aging ethnic minorities in India general developing and developed country in particular.

Frailty has long been deemed to be an inevitable consequence of ageing while some have thought of frailty as a state of pre-death. Many have used this concept synonymously with disability and co-morbidity (Pope & Tarlov, 1991; Hoffman, Rice, & Sung, 1996). There is clearly no consensus on the definition of frailty. Generally, the frailty has been emerging as a controversial and enigmatic concept. The emergence of frailty has an increasingly importance because of its prompts inquiry into what it is? Is it a disease? Is it just ageing? Or is it psychological vulnerabilities in the person? There are certain cases where some younger adults are also deemed as frail and some older ones are not.

In his study Gillick (2001) observed that "frailty is a syndrome in desperate need of description and analysis". Yet, many papers have highlighted the importance *of frailty in its relation to adverse outcomes which include dependency, disability, institutionalization, falls, injuries, acute illness, hospitalization, slow or blocked recovery and mortality* (Hogan, MacKnight, & Bergman 2003). Its high frequency decrease body's resiliency and its ability to compensate for environmental stressor during age related physiological changes. Life expectancies for men and women are increasing (U.S. Bureau of the Census, 1997), and over the age of 65 year the frailty has been projected to double in size by the year 2020 (Bokovoy & Blair, 1994) and over will be quadrupled (Zedlewski, Barnes, Burt, McBride, & Meyer, 1990). Unprecedented degree of population growth it continues in fashionable way it will have its serious repercussion on (U.S. Bureau of the Census, 1992) the elderly well being.

The elderly population is growing rapidly both in number and as a proportion of the total population in developing and developed countries alike. By the year 2020, one of every five residents of developed countries will be aged 60 year and older as well in the residents of developing nations. Older women greatly out-number older men in most of the nations across the world. Women from the great majority around 80 and over in most countries in the world are growing/steadily. The proportions of older women who are widowed are many times greater than the men particularly in all developed and developing nations.

The increasing weight of the elderly population, particularly in India has its own reasons. The improvement in life expectancy, decreased mortality rates due to improvement in medical sciences and migration of younger people to other countries to seek education and jobs and effects of natural disaster are some of them. In spite of govt. efforts to support increasing number of older dependents and to improve the quality of life care, the coverage has remained much smaller in terms of protected persons and levels of benefits provided. Breaking of joint family to nuclear one is also causing mental harassment to the elderly people. The demographic and socio-economic projections have further given rise to serious impact on ageing. The elderly in our country are considered to be a huge non-work force as the concept of forced retirement has a dual negative psycho-social and cultural effect both on the elderly as well as on the younger persons. Education in this respect can prove as a significant tool for changing their health, behaviour and cognitive process or personality development. Beside this home security, emotional security, vocational; surety etc. can further prove useful.

Politics of Aging

Aging is the accumulation of changes in an organism or object over time (Bowen & Atwood, 2004). It exerts its influence on humans being multifariously especially through physical, psychological, and social-cultural routes. Some dimensions of ageing have grown and expanded over time, whereas others declined significantly. Reaction time, for example, may appears slow with age, whereas the knowledge of world events and wisdom may expand with increasing age. Researches have shown that even late life potential exists for physical, mental, social growth and development. Thus, the ageing is an important part of all human societies reflecting the biological changes that occur, but also reflecting cultural and societal conventions.

According to a conservative estimate roughly there are 100,000 people worldwide those die each day of age-related causes (Aubrey & de Grey, 2007). The term "ageing" is somewhat ambiguous in nature need thorough understanding. Distinction is required between "universal ageing" (age changes that all people share), "probabilistic ageing" (age changes that may happen to some, but not all people as they grow older, such as the onset of type two diabetes) as well as psychological aging. The chronological ageing, refers to how old a person is but the most straightforward definition of ageing may be distinguished from "social ageing" (society's expectations of how people should act as they grow older) and "biological ageing" (an organism's physical state as it ages). There is also a distinction between "proximal ageing" (age-based effects that come about because of factors in the recent past) and "distal ageing" (age-based differences that can be traced back to a cause early in person's life, such as childhood poliomyelitis) Ian, 2006). Divisions are sometimes made between the young old (65-74) years, the middle old (75-84) years and the oldest old (85+) years.

In biology, *senescence* is the state or process of ageing. Cellular senescence is a phenomenon where isolated cells demonstrate a limited ability to divide in culture (the Hayflick Limit, discovered by Leonard Hayflick in 1961), while *organismal senescence* is the ageing of organisms. After a period of near perfect renewal (in humans, between 20 and 35 years of age), organismal senescence is characterized by the declining ability to respond to the stress, increasing homeostatic imbalance and increased risk of disease. This currently irreversible series of changes inevitably ends in death. Some researchers (specifically bio-gerontologists) are treating ageing as a disease as genes that have an effect on ageing. It is regarded in a similar fashion to other genetically influenced "conditions", potentially "treatable." Indeed, ageing is not an unavoidable property of life. Instead, it is the result of a genetic program. Numerous species shows that there are very low signs of ageing ("negligible senescence"), the best known being trees like the bristlecone pine (however Dr. Hayflick states that the Bristlecone Pine has no cells older than 30 years), fish like the Sturgeon and the Rockfish, Invertebrates like the Quahog and sea Anemone and Lobster.

In humans and other animals, cellular senescence has been attributed to the shortening of telomeres with each cell cycle; when telomeres become too short, the cells die. The length of telomeres is therefore the "molecular clock," predicted by Hayflick. Telomere length is maintained in immortal cells (e.g. germ cells and keratinocyte stem cells, but not other skin cell types) by the telomerase enzyme. In the laboratory, mortal cell lines can be immortalized by the activation of their telomerase gene, present in all cells but active in few cell types. Other genes are known to affect the ageing process; the sirtuin families of genes have been shown to have a significant effect on the lifespan of yeast and nematodes. Over-expression of the RAS2 gene increases lifespan in yeast substantially that has been found as quite important. In addition to genetic ties to lifespan, diet has been shown to substantially affect lifespan in many animals. Specifically, caloric restriction (that is, restricting calories to 30-50% less than an adlibitum animal would consume, while still maintaining proper nutrient intake), has been shown to increase lifespan in mice up to 50%.

In the USA, adulthood legally begins at the age of eighteen, while old age is considered to begin at the age of legal retirement (approximately 65). *The Juvenile* (via *Infancy /*

Childhood / Adolescence (Teenager)) stage is considered between 0 – 19 year. *The early adulthood stage* between 20 – 39 year followed with middle adulthood stage that is near begins near about 40 to 59 year of age. Similarly the *Late adulthood stage* appears between 60 to 79 year of age and the *Ancient stage i.e.* also known as Death / *Post-death*: Decomposition stage begins at 80 – 99 year. The age also be divided by decade basis 10 to 19 year onward the person is called as *Denarian*; from 20 to 29 year as *Vicenarian*; from 30 to 39 year as *Tricenarian;* 40 to 49 year as *Quadragenarian*; 50 to 59 year as *Quinquagenarian*; 60 to 69 year as *Sexagenarian;* from 70 to 79 year as *Septuagenarian*; from 80 to 89 year as *Octogenarian* from 90 to 99 year as *Nonagenarian;* 100 to 109 year as *Centenarianand* 110 years and older as *Supercentenarian.* People from 13 to 19 years of age are also known as ***teens*** *or* ***teenagers***. East Asian age reckoning is different from that found in Western culture. Traditional Chinese culture uses a different ageing method, called *Xusui* with respect to common ageing which is called *Zhousui.*

According to Sigmund Freud the focal objects for the developing child's energy serves to define five main stages of psychological development named *as oral* (0-18 months); *anal* (18 months - 3 1/2 years); *phallic* (3 1/2 years - 6 years) ;*latency* (6 years - puberty) and *genital* (puberty - adulthood). A newborn baby, according to Freud, is bubbling with energy (libido; psychic energy). However, this energy is without focus or direction, which would not allow for survival. How, then, does the child develop the ability to control and direct his/her energy? Psychic energy is an important concept in Freudian psychology. The structure of the mind and development all revolve around how the individual attempts to deal with psychic energy. Raw libidinal impulses provide the basic fuel that the mind runs on. But the vehicle (mind) needs to well-formed and well-tuned in order to get maximum energy. In order to understand development (and neuroses), then, we should? Follow the energy? And see where it goes. As with physical energy, psychic energy cannot be created or destroyed in a big picture sense; however it may dealt with non-obvious ways. So, where does the infants? Then the childs, the adolescent's, and adult's energy get focused? Freud believed that as development occurs the baby begins to focus on first one object than another. As the infant's focus shifts the style and type of gratification being sought changes. The life and works of Sigmund Freud deals with explanations of levels of consciousness, libido, id, ego, and superego, defense mechanisms, psychosexual stages of development, and repression.

Erickson a student of Freud explained eight stages of psychosocial development through which a healthy developing human should pass from infancy to late adulthood. In each stage the person confronts, and hopefully masters, new challenges. Each stage builds on the successful completion of earlier stages. The challenges of stages not successfully completed may be expected to reappear as problems in the future. First stage deals with hope between 0 to 1 year where basic needs are met by the parents. The second stage deals with 2 to 3 years where the child explores his surrounding; third stage with purpose that begins at 4 to 6 years where the child master the world and perform activities; fourth stage effect competence between 7 to 11years where the child is productive and wish to play with other; fifth stage deals with fidelity that involves 12 to 19 years where ego develop and child think how he appears to other. After this love stage begins with 20 to 34 years of age where the intimacy are developed ; the care stage begin from 35 to 65 years where the person guides and dictates other and wisdom stage comes after 65 onward here the person becomes a senior citizen who tends to slow down activities and explored life as a retired person. Piaget claims that the idea that cognitive development is at the centre of human organism and language is contingent on cognitive development. He gave four stages of cognitive development named as: *Sensorimotor stage* where infants construct an understanding of the world by coordinating sensory experiences. Second is *preoperational stage* where the child observes sequences of play. Third is concrete operational stage where the child learns appropriate use of logicand Formal operational stage where the child thinks abstractly and more logically.

According into Kohlberg theory of development, the moral reasoning is the basis for ethical behavior. He identified six developmental stages, each more adequate at responding to moral dilemmas than its predecessor. Kohlberg's six stages can be more generally grouped into three levels of two stages each: pre-conventional, conventional and post-conventional. Kohlberg followed the development of moral judgment far beyond the ages studied earlier by Piaget, who also claimed that logic and morality develop through constructive stages. Expanding on Piaget's work, Kohlberg determined that the process of moral development was principally concerned with justice, and that it continued throughout the individual's lifetime, a notion that spawned dialogue on the philosophical implications of such research. He relied for his studies on stories such as Heinz dilemma, and was interested in how individuals would justify their actions if placed in similar moral dilemmas. He then analyzed the form of moral reasoning displayed, rather than its conclusion, and classified it as belonging to one of six distinct stages.

Most legal systems define a specific age for an individual when he is allowed or obliged to do something. These ages include *voting age, drinking age, age of consent, age of majority, age of criminal responsibility, marriageable age, age of candidacy, and mandatory retirement age.* Admission to a movie for instance, may depend on age according to a motion picture rating system. A bus fare might be discounted for the young or old. The dependency theory or dependencia theory predicated on the notion that resources flow from a "periphery" of poor and underdeveloped states to a "core" of wealthy states, enriching the latter at the expense of the former. It is a central contention of dependency theory that poor states are impoverished and rich ones enriched by the way poor states are integrated into the "world system." Vygotsky focused on the connections between people and their socio-cultural context in which they act and interact in shared experiences (Crawford, 1996). According to Vygotsky, humans use tools that develop from a culture, such as speech and writing, to mediate their social environments. Initially children develop these tools to serve solely as social functions, ways to communicate needs. Vygotsky believed that the internalization of these tools led to higher thinking skills. Social development theory argues that the social interaction precedes development; consciousness and cognition is the end product of socialization and social behavior.

Generally the old age (also referred to as one's eld) consists of ages nearing or surpassing the average life span of human beings, and thus the end of the human life cycle. The terms for old people include *seniors* (American usage), *senior citizens* (British and American usage) and the *elderly.*

As occurs with almost any definable group of humanity, some people will hold a prejudice against others - in this case, against old people. The social definitions of being considered "child, teen, young adult, adult, middle-aged, elderly, old, senior citizen, ancient" are culturally variable but most societies have an implied chronological age at which such labels are used. Age life cycle is composed of the total average life span pitted against the individual's life span, age specific societal restrictions, the individuals age related self-perceptions and the prevalence of age related labeling in a given society (Dixon, 2001). The medical study of the aging process is gerontology, and the study of diseases that afflict the elderly as geriatrics. The boundary between middle age and old age cannot be defined exactly because it does not have the same meaning in all the societies. There is often a general physical decline, and people become less active. There are wrinkles and liver spots on the skin ; change of hair color to gray or white; hair loss, lessened hearing; diminished or poor eyesight , slower reaction times and agility , reduced ability to think clearly , difficulty is recalling or memories , lessening or cessation of sex, erectile dysfunctioning in men, and abdominal/vaginal pain in women as well as but often simply a decline in libido and greater susceptibility to bone diseases such as osteoarthritis, and depletion of bone marrow.

In the United States the proportion of people aged 65 years or older increased from 4% in 1900 to about 12% in 2000. In 1900, only about 3 million of the nation's citizens had reached to 65 years. By the year 2000, the number of senior citizens had increased to about 35 million. Population experts estimate that more than 50 million Americans - about 17 percent of the population - will be 65 years or older upto 2020. In most parts of the world, women live, on average, longer than men. According to Erik Erikson's "eight stages of life" theory, the human personality is developed in a series of eight stages that take place from the time of birth and continue throughout an individual's complete life. He characterized old age as a period of "Integrity vs. Despair", during which a person focuses on reflecting back on their life. Those who are unsuccessful during this phase will feel that their life has been wasted and will experience many regrets.

There are some people who become very famous in old age. For example a lady named Ann Nixon Cooper, who at age 106 years made national news during the 2008 US presidential election for voting for Barack Obama. Granny D, political activist who ran for public office at the age of 94 years; James Fisher, blacksmith who returned from retirement to become the first person over the age of 100 years to achieve the ACA qualification; Enrico Dandolo, who led the infamous Fourth Crusade in his year 80s; Sadie and Bessie Delany, civil rights activists; Ruth Ellis, 101-year-old African-American LGBT activist; Florence Holway, rape survivor and activist ; Mary Harris "Mother" Jones, Irish-American labor organizer; Maggie Kuhn, activist and founder of the Gray Panthers; Mae Laborde, actress who began acting in her 90s; Bernando LaPallo, American doctor who wrote his first book at age 107 years and is the worlds oldest blogger at age 109 years; Buster Martin, said to be the oldest worker in the UK at age 103 years ; Grandma Moses, American folk artist ; Narses, who became a successful general at 74 years; Peter Oakley, aka geriatric1927, British senior famous for his YouTube videos; and Nola Ochs, became the oldest college Beside this Clara Peller, Emily Perry, Mary Jane Rathbun, Malvina Reynolds, Olive Riley, and Arthur Winston are some other name. In India Mahatma

Gandhi and Pt. Jawahar Lal Nehru and many other leaders and eminent personalities become very famous in old age even today are memorable.

Modernization and age stratification theories run into the problem of aggregate data obscuring within group differences even within one culture let alone in a cross cultural context. Old people are independently wealthy and educationally privileged above the average citizen. Sometime family is seen in the less developed nations to pick up the slack and provide the older generation a safety net whereas the older people in the modern society are seen as more wealthy and self-sufficient but isolated from their kin. Specification of the work life cycle involvement is needed to be able to make any categorical claims on the relationship to the means of subsistence on group level. Some life cycle involvement is necessary for physical, social and psychological survival. The unit analysis for exploring web of life cycle involves family, education, work and leisure can be individual, group or society.

Life Style of Elderly Frail People

The life-style is a way of life for an individual or community that is particularly expressed through interests and opinions. The life course approach provides an attractive framework towards understanding frailty and its determinants. It integrates biological, social, clinical, cognitive, psychological and environmental factors; those interact across a person's life span. Studies on the risk factors for frailty support this approach. The researchers focused on what people can do in their early elderly years to live longer while maintaining their good health and physical functioning is a vital issue as the population ages. Better life style depends upon one daily activities or habits e.g. yogic exercises, not to smoke, drink, maintained normal body weight through normal eating, control blood pressure by thinking positive and eating nutritious food, and as well as controlling diabetes by applying better life style. The life styles in aging deals with ones cycles of family, education, work, leisure and better quality of life will be in terms of physical, social and mental well-being with carryover effect in old age.

Lifestyle changes are the hardest ones to make. Most people would say that they don't want to have extra years added to their life if those years are going to be ones of disability and disease. The psychology of how people engage with healthy habits, and maintain their engagement in healthy lifestyles is thus a vital facet of any attempt by countries and corporate to change the awful effects of modern western existence on body and brain health. Effective measures have been applied for reducing mortality and morbidity in the developed world. However, despite decades of health promotion, there has been no significant difference to lifestyles and instead there are rising levels of inactivity and obesity. The role of psychology in the field of elderly wellbeing can play in reversing the trend to the deleterious lifestyle choices. Health promotion has taught people what a healthy lifestyle. More appropriately it focuses on smoking, eating, physical activity, drinking, sex and drug use - as well as combinations of behaviors.

According to both Individual Psychology and Interpersonal Theory, individuals develop patterns early in life through interactions with their families, most importantly with their parents. Adler described the patterns that individuals follow through their lives as their lifestyle (Sweeney, 1975). A lifestyle, according to Individual Psychology, is the characteristic way in which individuals learn in their families of origin to pursue social significance in interpersonal relationships. Similarly, Interpersonal Theory depicts individual development as an interpersonal process in which he learn to avoid or manage anxiety and maintain self-esteem in interpersonal relationships (Sullivan, 1953; Teyber, 2000). Painful anxiety arises when individuals anticipate rejection by parents, others, and oneself (Grey, 1988). Through repeated interactions, a elderly forms a self-other relational pattern designed to minimize or avoid anxiety and maintain self-esteem (Sullivan).

Both Individual Psychology and Interpersonal Theory suggest that the individual develop patterns of interaction to function in their social (family) environments and that the style of interaction continues when they become adults. Furthermore, both theories suggest that this interaction style could impede social or interpersonal functioning. Individual Psychology and Interpersonal Theory assume that humans are social beings, and human behavior can only be understood as a function of individuals' perception of their social context (Shilling, 1984; Sherman & Dinkmeyer, 1987). Adler held a holistic and phenomenological perspective on human experience (Sweeney, 1981). Similar to Individual Psychology, Interpersonal Theory offers a holistic view of humans acting as an organism or personality (Mullahy, 1970) and views behavior as "inclusive of the subjective and behavioral activities of all participants" (p. 556), as opposed to outside the individual (Grey, 1988). Individual Psychology posits that social significance is productively achieved through the development of social interest that contributes better to the society (Sweeney, 1975). It also views social interest as the criterion for mental health, as opposed to the feelings of inferiority leading to the need for self-protection

(Ansbacher, 1991). According to Adler (1964) "all neurotic Symptoms have as their object the task of safeguarding the patient's self-esteem" (p. 263). Sullivan (1956) had a similar view regarding the fundamental importance of interpersonal relations to well-being. Sullivan (1953) stated that "one achieves mental health to the extent that one becomes aware of one's interpersonal relations" (p. 26), and uses that information to improve one's relationships.

There are series of risk factors in mid to late life (including cognitive impairment, depression, disease burden, increased/ decreased BMI, lower extremity function limitation, decreased social contacts, low physical activity, alcohol consumption, poor perceived health, smoking and vision impairment) for functional decline. There is evidence for an association among biological, psychological and social risk factors those

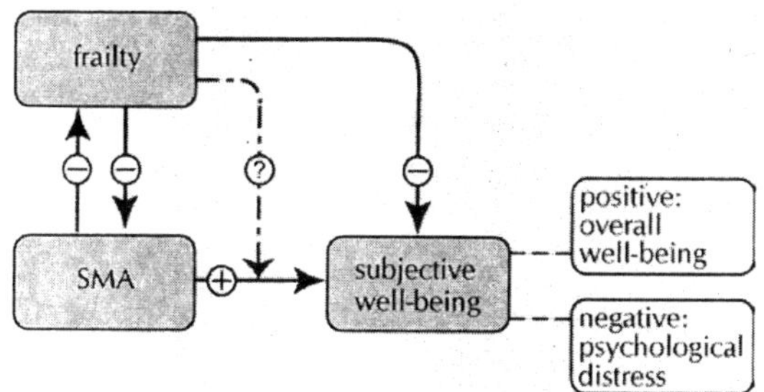

Figure 1 The expected influence of frailty and Self management of aging (SMA) on wellbeing.

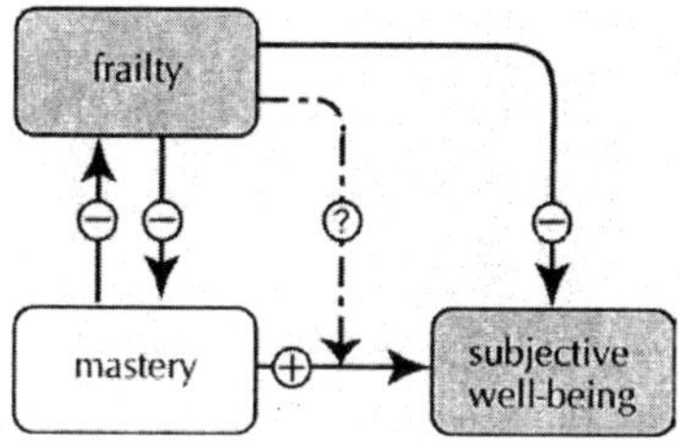

Figure 2 The expected influence of mastery on wellbeing.

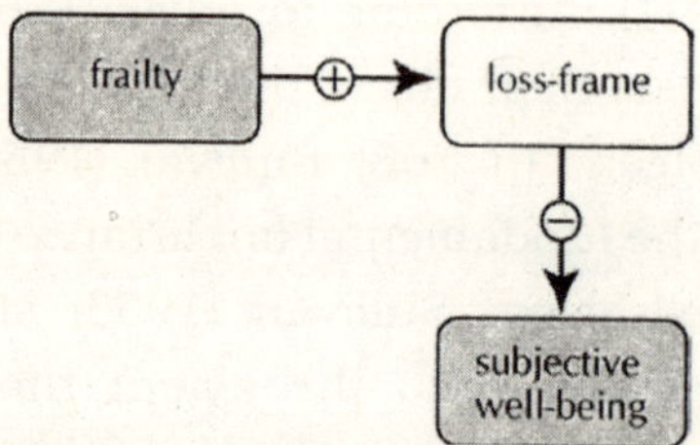

Figure: 3 The expected influence of frailty via a loss frame on well-being.

exert influence on the health. It is generally associated with the onset of frailty and its adverse outcomes. Frailty is a dynamic and complex process with multiple interacting components. Observational studies on aging suggest associations between several lifestyle factors (exercise, nutrition, education, socioeconomic status, social/ intellectual activities) and the onset of frailty. These findings provide opportunities for the development of interventions to promote healthy ageing, reduce the incidence of frailty, delay its onset and/or reduce the number of years in dependency (Vita, Terry, Hubert, & Fries, 1998)

The quality of life, dignity and personal well-being must always be the focus of the care and support someone receives. Social care needs to focus on outcomes - people may receive care which meets their physical needs, but their emotional and mental wellbeing are too often overlooked completely. In some care homes there are little understanding of dementia or depression, meaning that both conditions go unrecognized and untreated. The emotional exhaustion, is believed to be at the core of burnout (Maslach, 1982), and is characterized by a lack of energy and a feeling that one's emotional resources are used up (Cordes & Doughtery, 1993). Researchers have linked burnout to a variety of mental and physical health problems (Burke & Deszca, 1986; Maslach

& Pines, 1977), including decreased self-esteem, increased levels of irritability, depression, and anxiety, as well as the deterioration of family and social relationships (Jackson & Schuler, 1983) that in turn can be associated with psychological frailness. .

According to American Medical Association one of the most important tasks that the medical community faces today is to prepare for the problems in caring for the elderly in the 1990s and the early 21st century (AMA, 1990). The report particularly emphasized the growing population of "frail and vulnerable older adults". The association estimated that in the community, 10 to 25% of those above 65 years are frail and 46% of those above 85 years are also suffering with frail. The proportion of older persons above the age of 65 was in Singapore was 6.8% found frail and in 1995. It is projected to increase to 20% by the year 2030(The National Survey of Senior Citizens in Singapore, 1995). The increase in life expectancy and better healthcare would introduce an emergence of the very old and frail populations. Understanding the underlying mechanisms of the frail state points to opportunities for health promotion, prevention and interventions aimed at either delaying the onset of frailty or reducing its adverse outcomes.

The concept and history of frailty dates back to the 1970s with the endeavor of Monsignor Charles Fahey and the Federal Council on Aging in the United States who were credited with introducing the "frail elderly" to describe a particular segment of the population (Maddox, 1987 ; Albert, 1995) . This happened when the heterogeneity of the older population became more accepted. The term was selected to focus attention on a group of elderly with physical debilities, emotional impairments and debilitating social and physical environments. It was used as an administrative device for triggering access to core support services for the elderly. In the 1980s, researchers began to explain the term. Early definitions would include those aged 75 and older; a vulnerable population of seniors due to physical and mental impairment; individuals admitted to a geriatric programme; institutionalization; and those dependent on others for activities of daily living. (Streib, 1983; Reid, Gallagher & Bosworth, 1986; Williams, Wynne, Woodhouse, & Rawlins, 1989). Chronic disease and its sequelae were felt to be the cause of increased vulnerability and frailty (Woodhouse, Wynne, Baillie, James, & Rawlins, 1998). For the human brain, there's no such thing as over the hill. Psychologists researching the normal changes of aging have found that although some aspects of memory and processing change as people get older, simple behavior changes can help people in staying sharp for as long as possible.

Quality of Life among Rural and Urban Elderly Frail- People

According to World Health Organization (2001), the "quality of life is as an individuals perception of their position in life in the context of the culture and value system in which they live and in relation to their goal, expectations, standard and concerns. It is a broad ranging concept affected in complex way by the persons physical health, psychological status, level of independence, social relationship and their relationship to salient features of their environment (Bowling, 1995, p. 3). In India more than 76% population as well as in India approximately 90% population lives in rural areas. The quality of life the people that involves satisfying and satisfactory condition while living in rural areas in past were with full of obstacles or difficulties as compared to their urban counterpart.

Due to various landform and topology the life of the people especially on mountainous Himalaya were quite difficult. Beside these religious disparities since the time memorial has remained there those have affected the people from the mind and body. The scheduled castes and scheduled tribes since time immemorial have remained in the malicious grip of poverty. Be it a child, youth or elderly people belonging to these segments suffers from prolonged deprivation thereof enmeshed in psychopathology. The less land holding push them towards bad habit such as drinking and smoking that in turn make them psychologically frail. For getting rid of their pathological conditions the people in turn prefer to consume alcohol and other substance for getting rid of their tension that in turn again make them vulnerable towards pathology.

In order to gain insight into the relation between health and people's environment, literature published between 1985 and 1994 was gathered from several international databases. First, emphasis in past research has been primarily on urban constraints rather than opportunities. Positive aspects of urban living are often insufficiently appreciated. Second, positive and negative environmental aspects have an effect on health that is often dependent on individual characteristics. The extent to which the environment exerts influence on a person's health is dependent on that person's individual characteristics. Large scale individual based longitudinal data should be studied in order to gain more insight into the relative importance of the geographical drift hypothesis (Verheij, 1996).

Frailty and disability were often used interchangeably. There were a growing number of researchers who postulated that frailty was another term for disability in the elderly. By the 1990s, researchers felt that equating frailty with disability was inadequate. A number of difficult questions were asked. Were all older disabled patients frail? If not, why weren't they? How did one end up frail? Was it inevitable? What were the underlying mechanisms? Was frailty preventable? Is it clinically advantageous to delineate frailty? Current research is attempting to delineate the components of frailty (how do we identify it?) and the underlying physiological and biological mechanisms important in the development of a frail state.

Research has focused in particular on memory and ageing, and has found decline in many types of memory with ageing, but not in semantic memory or general knowledge such as vocabulary definitions, which typically increases or remains steady. Early studies on changes in cognition with age generally found declines in intelligence in the elderly, but studies were cross-sectional rather than longitudinal and thus results may be an artifact of cohort rather than a true example of decline. Intelligence may decline with age, though the rate may vary depending on the type, and may in fact remain steady throughout most of the lifespan, dropping suddenly only as people near the end of their lives. Women are more vulnerable to higher levels of loneliness because of their greater longevity compared to men. As women age outlive spouses, friends and family members who previously provided the social and emotional support that are important for health and wellbeing (Havens & Hall, 2001, p.130)

An American Medical Associations white paper concluded that "one of the most important tasks that the medical community faces today is to prepare for the problems in caring for the elderly in the 1990s and the early 21st century". The report particularly emphasized on the growing population of "frail, vulnerable older adults". It estimated that in the community, 10 to 25% of those above 65 years are frail and 46% of those above 85 years are frail. In Singapore, the medical community is also facing with similar problems in caring for an ageing population. The proportion of older persons above the age of 65 was 6.8% of the population in 1995 and is projected to increase to 20% by the year 2030. The increase in life expectancy and better healthcare would introduce an emergence of the very old and frail populations. Understanding the underlying mechanisms of the frail state points to opportunities for health promotion, prevention and interventions aimed at either delaying the onset of frailty or reducing its adverse outcomes.

PERSPECTIVES OF PSYCHOLOGICAL FRAILNESS

The only standard we have for judging all of our social, economic, and political institutions and arrangements as just or unjust, as good or bad, as better or worse, derives from our conception of the good life for man on earth, and from our conviction that, given certain external conditions, it is possible for men to make good lives for themselves by their own efforts.

—Mortimer J. Adler

The earth is only the planet in the universe that is very well known to harbor life on it. Evidence suggests that the life on the earth existed about 3.7 billion years ago. A creation myth is a supernatural story that even describes the beginnings of "universe" (cosmogony), "earth" as well as the "humanity" or the "life" often as a deliberate act by one or more deities.The origin of life attempt to find a mechanism that explains about the formation of a primordial single cell organism from which all life originated. The idea that the "Earth" is alive was probably as old as humankind was. A Scottish scientist, James Hutton in 1785, the father of "Geology", propounded it by stating that the earth once was a super organism. His idea of living earth later vanishes in the intense reductionism of the nineteenth century.

Similarly, the Gaia hypothesis as proposed by scientist James Lovelock in 1960 had explored the idea that the life on earth functions as a single organism that in turn defines and maintains environmental conditions that has prime importance in the survival. There appeared different hypotheses those ranges from simple organic molecules via pre-cellular life to the proto-cells and metabolism. From infancy to old age, perhaps, the superpower that we recognized in the form of "God" or goddess" might have played role in shaping physiology, psychology and socio-cultural processes of the human being.

Research into ageing has traditionally been growing in discreet areas, looking at either psychological issues, community influences or health related issues. These factors combine with personal ability factors and then we make plan for the future. The original research actually started back in the 1980s when people's cognitive abilities were explored when they get older. The finding revealed that the age has little bearing on cognitive performance and there's great variability amongst older people as well. In the socio-environmental model of health, each of these separate conceptual domains was recognized.

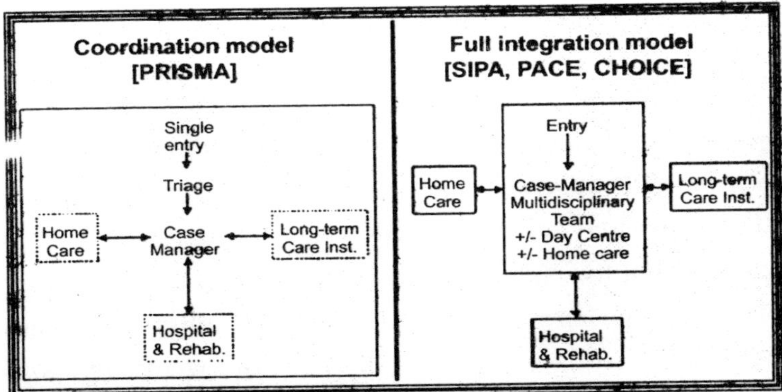

Figure 2.2.1 Comparison of two model of health care delivery.

Therefore the health encompasses complex multi-dimensional factors that come in contact with cultural, environmental and psycho-social influences. The limitations of the "biomedical" paradigm of health have been recognized, principally that this model only deals with disease. Consequently, any measure of health needs to assess social and emotional aspects of health as well as assessing presence or absence of disease (Wilson & Cleary, 1995). Sometimes factors like low income, reduced physical mobility due to illness, or lack of appropriate transport can leave some people at home alone with nothing much to do. But there are lots of ways to stay involved in the wider community.

Perspectives of Psychological Frailness

The contemporary perspectives on frailty should not be viewed as mutually exclusive. The diversity in concepts has its duplication and overlap between models. It also speaks to the isolation of researchers working on frailty (Hogan, MacKnight, & Bergman, 2003). A simple model for frailty states that it is an intrinsic to ageing. The ageing has been defined as a process resulting in diminished reserves in most physiologic, psychological, sociological and environmental adjustment related systems and exponentially increasing vulnerability to most diseases and death. However, while some adults are more likely to be frail with increasing age, frailty is not a state universally present in all older persons. This suggests that frailty is associated with ageing but not an inevitable consequence.

Older adults vary in their overall robustness (Verbrugge, 1991). Biomarkers of ageing (including frailty), that were deemed inevitable, e.g. grey hair, skin wrinkles, poor endurance, work output, are not only influenced by ageing but also by various genetic and environmental factors and no two individual ages in the same way (Miller, 2003). Lipsitz likens the changing view of frailty to the change in the way Alzheimer's disease was perceived. "Thirty years age, Alzheimer's was thought to be an inevitable consequence of aging; we called it senility. The same is true of frailty that it is a syndrome that probably involves some physiological underpinnings (Kolata , 2004). It has also been suggested that frailty is synonymous with accelerated or pathological ageing as opposed to successful or healthy ageing (van Staveren, de Graaf, & De Goot, 2002).

Most often the frailty has been proposed to be a result of concurrent impairments in multiple biologic systems. Underlying mechanisms have been suggested which include impaired strength, balance and endurance. There is an attempt to link the patho-physiology and clinical manifestation of frailty. The frailty generally is considered to be related to the patho- physiological effects of an altered metabolic balance, manifested by cytokine over-expression and hormonal decline. The "cycle of frailty" involves sarcopaenia, neuroendocrine dysregulation and immune dysfunction being the "physiological triad" to the frailty. Although a variety of physiologic underpinnings have been proposed, most researchers would agree that frailty results from a multi-system reduction in reserve capacity to the extent that a number of physiological systems are close to the threshold of symptomatic clinical failure.

The demographers have developed models for survival to include a factor for unobservable heterogeneity or frailty (Suchindran & Koo, 1997). The concept of frailty has been used to explore susceptibility to the ageing process as a whole (Giard, Lichenstein & Yashin , 2002). This susceptibility may reflect genetic pre-disposition or cumulative effect of environmental exposures. Another demographic observation was that death rates decrease with age for humans as well as a host of other organisms (Vaupel, Carey, Christensen,

Johnson, Yashin, & Holm, et al, 1998). A possible explanation for this is the heterogeneity of frailty. The frail individuals are more likely to die and leave as a selected subset of robust individuals (Hitt, Young-Xu, Silver & Perls, 1999). It has also been postulated that the mortality plateau occurs because of the slowing of the ageing process at the level of the individual (Masoro, & Austad, 2001). A mathematical model can be defined as a symbolic device utilizing mathematical reasoning to simulate and predict aspects of behavior of a system. There are advantages in elucidating concepts in this model as it forces the researcher to articulate a clear understanding of the assumptions underlying their models.

Certain scholars have linked chaos theory to explore the dimension of frailty. A loss of complexity in the dynamics of numerous physiological systems (heart rate variability, blood pressure, hormonal rhythms, and postural sway) reduces an individual's ability to adapt to stress and that in turn leads to the syndrome of frailty. The output of these processes demonstrates complex variability that can be quantified using the concept of fractals, derived from the field of nonlinear dynamics (Lipsitz & Goldberger, 1992; Lipsitz, 2004). Mitnitski has calculated a frailty index, a collection of 40 variables conceptualized similarly to the accumulation of biologic burden in allostatic load. He postulated that the approach emphasizes the "integration" (or lack of integration) of many components in a complex system (Mitnitski, Song, & Rockwood, 2002).

The ageing is a process of gradual accumulation of damage in cellar tissues of the body leading eventually to frailty and an increased risk from a spectrum of age-associated diseases. Mechanisms leading to cellular damage include cumulative damage to mitochondrial and nuclear

DNA (Kirkwood, 1989). These mutations lead to a range of bioenergetically deficient cells which characterize a main phenotype of frailty: muscle weakness and sarcopaenia (Linnane, Marzuki , Ozawa, & Tanaka , 1989). The term sarcopaenia (from the Greek word sarx for flesh and penia for loss) was first coined by Rosenberg to identify loss of muscle mass and function (Rosenberg, 1993). It has been suggested as the major underlying cause of frailty (Pendergast, Fisher, & Calkins, 1993; Roubenoff, & Harris, 1997). The consequences include decreased strength, metabolic rate and maximal oxygen consumption. These physiologic decrements contribute to weakness and a loss of independence (Dutta, 1997). A cross-sectional study found that sarcopaenia was associated with increased functional impairment and disability. However, its underlying mechanism is not fully understood. The mechanisms proposed have included hormonal dysregulation, oxidative injury from the plasma thiol/disulfide redox state and nutrition.

With ageing, there is a decline in the functioning of a number of endocrine systems. This leads to the development of menopause/andropause, adrenopause and somatopause (Lamberts, van den Beld , & van der Lely , 1997). These hormonal deficiencies may contribute to frailty. In the late 19th century, Brown-Sequard (in his 70s), treated his declining endurance, strength and mental abilities with crude testicular extract derived from dogs and Guinea pigs (Aminoff, 2000). Suppression of insulin-like peptides, insulin-like growth factor, lipophilic signalling molecules, and sterols or their receptors can delay age-related functional decline and increase stress resistance in animal models. Hormone's important to the development of frailty that have been proposed include testosterone, leutinising hormone and cortisol. A low level of chronic inflammation secondary to age-related deregulation of the immune system has been proposed (Cohen, 2000; Cohen, Harris & Pieper, 2003). Potential markers include reactive proteins, tumor necrosis factor-alpha and interleukin-6.

With the life course approach, long term effects of biological, behavioral and/or social pathways on risks of developing a disease, condition or state are studied during gestation, childhood, adolescence, young adulthood and later adult life. *The person with psychological frailness shows allostatic load that involves "wear and tear" or "use it and lose it" that occurs in an organism over time to maintain a steady state.* It is conceptualized as the cumulative biological burden exacted on the body through attempts to adapt to life's demands (Seeman, Singer, Rowe, HorwitzI, & McEwen, 1997).The life course postulated by Bortz considers frailty as a result of lessened load that leads to linked and parallel loss in form and function. A simplification of this concept is "use it or lose it". Allostatic load and symmophosis are not mutually exclusive concepts rather is an optimal balance between the two that later captured the concept of hormesis. It is known that a number of mild stressors have been found to increase longevity in animal models (Parsons, 2002). Hormesis might result from either direct stimulation or over-compensatory response to noxious stimuli (Calabrese & Baldwin, 1985). To prevent frailty then, the individual has to be exposed to a mild load or mental stressors in order to induce the appropriate adaptive response.

There have been attempts to develop frailty models that incorporate a diverse number of factors operating at different levels (molecular to functional levels) and represent its various dimensions. Brocklehurst offered a "model of breakdown of frailty with assets and deficits. The assets includes health, functional capacity, positive attitude to life, caregiver availability and other resources while the deficits proposed included chronic disease, disability, dependency and caregiver burden which threaten independence. *Frail individuals are those whose deficits outweigh assets* (Brocklehurst, 1985). Lebel, in his model of frailty included predisposing factors that influence the development of age-associated degenerative changes that can cause impairments. Cognition, neuro-locomotor and energy metabolism impairments were emphasized (Lebel, Leduc, & Leclerc, 1999).

There is growing consensus that frailty is a syndrome that can be identified and measured clinically. It is distinct from disability and co morbidity. It is a state of reduced homeostasis leading to increased vulnerability and risk of adverse outcomes. It is the result of the impact of multiple system impairment with critical changes in its reserve capacities, especially metabolic, cardiovascular, musculoskeletal, immunologic and neurological systems. The frailness represents a dynamic, complex interaction of biological, psychological, cognitive and social factors and a complex interplay of assets and deficits. Fried developed the "frailty phenotype", which proposes an operational definition to the clinical syndrome of frailty.

The parameters of frailty deals with self-reported and performance indicators of poor hand grip strength, slow walking speed, poor levels of physical activity, unintentional weight loss (4.5kg/year) and positive report of exhaustion/ fatigue followed with mental vulnerabilities. Individuals with three or more of these characteristics are classified as "frail" while those with one or two are labeled as "pre-frail". Using this definition, frailty has an independent predictive validity for mortality, disability and hospitalization (Fried, Tangen,

Waltson, Newman, Hirsch, & Gottdiener et al., 2001). Fried also hypothesized a cycle or spiral of frailty to unify the relationship of the components in the "frailty phenotype". The key components in this "phenotypic expression of the negative cycle" are chronic under nutrition, sarcopaenia and declines in strength, power and exercise tolerance with declines in activity and total energy expenditure. Thus, it is a core cycle of negative energy balance.

The researchers and clinicians have also considered working framework for studying frailty. The framework holds that biological, psychological, social and environmental factors that interact across the life course are the determinants of the onset of frailty. The frail people show decreased physical activity, weakness, and decreased endurance, slowness, under nutrition as wells cognitive, psychological and perhaps social components. The pathway from frailty to its adverse outcomes too is affected by various biological, psychological, social and societal modifiers. Many factors such as demographic (accelerated ageing of the population), social (break-up of families, children moving away to find work), economic (low income women living alone), health (increased life expectancy, high incidence of disabilities) and financial (reduced health care budgets)-are putting strong pressure on both the demand for and the supply of services for this clientele.

Functional decline generates an increased need, for both the dependent individuals and their families, for evaluation, treatment, rehabilitation, psychological and social support, help to remain at home, and temporary or permanent long-term care facilities. These multiple needs can also change quickly over time due to the biological, psychological and social vulnerability of this frail clientele. In terms of supply, a wide range of resources and services involving numerous practitioners and partners have been developed over the past twenty years to try to meet these needs. However, continuity-related problems compromise both service accessibility and the efficiency of health care services.

Biological Theories of Frailness

(I) **Genetic theory:** At present, the biological basis of ageing is unknown. Most scientists agree that substantial variability exists in the rates of ageing across different species, and that this to a large extent is genetically based. In model organisms and laboratory settings, researchers have been able to demonstrate that selected alterations in specific genes can extend the lifespan (quite substantially in nematodes, less so in fruit flies, and even less in mice). Nevertheless, even in the relatively simple organisms, the mechanism of ageing remains to be elucidated. Because the lifespan of even the simple lab mouse is around 3 years, very few experiments directly test specific ageing theories. It was US National Institute on Aging currently funds an intervention testing program; whereby investigators nominate compounds to have evaluated with respect to their effects on lifespan and age-related biomarkers in outbreed mice. Previous age-related testing in mammals has proved largely irreproducible, because of small numbers of animals, and lax mouse husbandry conditions. The intervention testing program aims to address this by conducting parallel experiments at three internationally recognized mouse ageing-centres, the Barshop Institute at UTHSCSA, the University of Michigan at Ann Arbor and the Jackson Laboratory. While the hypotheses below reflect some of the current direction in biological ageing research, none of them is accepted as a "theory" in the sense of the "theory of gravity" or "theory of relativity".

(II) **Misrepair-Accumulation Theory:** This very recent novel theory by Wang et al. suggests that ageing is the result of the accumulation of "Misrepair". Important in this theory is to distinguish among "damage" which means a newly emerging defect before any reparation has taken place, and "Misrepair" which describes the remaining defective structure after (incorrect) repair. According to this theory there is no original damage left unrepaired in a living being. If damage was left unrepaired a life threatening condition (such as bleeding, infection, or organ failure) would develop. Misrepair, the repair with less accuracy, does not happen accidentally. It is a necessary measure of the reparation system to achieve sufficiently quick reparation in situations of serious or repeated damage, to maintain the integrity and basic function of a structure, which is important for the survival of the living being. Hence the appearance of Misrepair increases the chance for the survival of individual, by which the individual can live at least up to the reproduction age, which is critically important for the survival of species. Therefore the misrepair mechanism was selected by nature due to its evolutionary advantage. However, since misrepair is a defective structure is invisible for the reparation system, it accumulates with time and causes gradually the disorganization of a structure (tissue, cell, or molecule); this is the actual source of ageing. Hence, ageing is the side-effect for survival, but important for species survival. Thus Misrepair might represent the mechanism by which organisms are not programmed to die but to survive (as long as possible), and ageing is just the price to be paid.

- ***Telomere Theory:*** It states that Telomeres (structures at the ends of chromosomes) have experimentally been shown to shorten with each successive cell division. Shortened telomeres activate a mechanism that prevents further cell multiplication. This may be an important mechanism of ageing in tissues like bone marrow and the arterial lining where active cell division is necessary. Importantly though, mice lacking telomerase enzyme do not show a dramatically reduced lifespan, as the simplest version of this theory would predict.
- ***Reproductive-Cell Cycle Theory:*** It states that and that ageing is regulated by reproductive hormones that act in an antagonistic pleiotropic manner via cell cycle signalling, promoting growth and development early in life in order to achieve reproduction, but later in life, in a futile attempt to maintain reproduction, become deregulated and drive senescence (dyosis).
- ***Wear-and-Tear Theory:*** It states that the very general idea that changes associated with ageing are the result of chance damage that accumulates over time.
- ***Somatic Mutation Theory:*** It states that the ageing results from damage to the genetic integrity of the body's cells.
- ***Error Accumulation Theory:*** It gives an idea that the ageing results from chance events that

escape proof reading mechanisms, which gradually damages the genetic code.

- ***The Viral Theory of Aging:*** highlights that the known causes of cancer (radiation, chemical and viral) account for about 30% of the total cancer burden and for about 30% of the total DNA damage. DNA damage causes the cells to stop dividing or induce apoptosis. DNA damage is thought to be the common pathway causing both cancer and aging. It seems unlikely that the estimates of the DNA damage due to radiation and chemical causes have been significantly underestimated. Viral infection would appear to be the most likely cause of the other 70% of DNA damage especially in cells that are not exposed to smoking and sun light.
- ***Evolutionary Theories:*** are speculative in nature; do not provide readily testable methods of confirmation.
- ***Accumulative-Waste Theory:*** of ageing points to a build-up of cells of waste products that presumably interferes with metabolism. Autoimmune Theory gives an idea that ageing results from an increase in auto-antibodies that attack the body's tissues. A number of diseases associated with ageing, such as atrophic gastritis and Hashimoto's thyroiditis, are probably autoimmune in this way. While inflammation is very much evident in old mammals, even SCID mice in SPF colonies still senescence.
- ***Ageing-Clock Theory:*** It emphatically visualizes ageing as the result of preprogrammed sequence,

as in a clock, built into the operation of the nervous or endocrine system of the body. In rapidly dividing cells the shortening of the telomeres would provide just such a clock. This idea is in direct contradiction with the evolutionary based theory of ageing. Cross-Linkage Theory suggests that that ageing results from accumulation of cross-linked compounds that interfere with normal cell function.

- ***Free-Radical Theory talks about:*** unstable and highly reactive organic molecules, also named reactive oxygen species or oxidative stress) create damage that gives rise to symptoms we recognize as ageing.
- ***A General Reliability Theory about:*** systems failure allows researchers to predict the age-related failure kinetics for a system of given architecture (reliability structure) and given reliability of its components. Reliability theory predicts that even those systems that are entirely composed of non-ageing elements (with a constant failure rate) will nevertheless deteriorate (fail more often) with age, if these systems are redundant in irreplaceable elements. Ageing, therefore, is a direct consequence of systems redundancy. Reliability theory also predicts the late-life mortality deceleration with subsequent leveling-off, as well as the late-life mortality plateaus, as an inevitable consequence of redundancy exhaustion at extreme old ages. The theory explains why mortality rates increase exponentially with age (the Gompertz law) in many species, by taking into account the initial flaws

(defects) in newly formed systems. It also explains why organisms "prefer" to die according to the Gompertz law, while technical devices usually fail according to the Weibull (power) law. Reliability theory allows specifying conditions when organisms die. According to the Weibull distribution, organisms should be relatively free of initial flaws and defects. The theory makes it possible to find a general failure law applicable to all adult and extreme old ages, where the Gompertz and the Weibull laws are just special cases of this more general failure law. The theory explains why relative differences in mortality rates of compared populations (within a given species) vanish with age (compensation law of mortality), and mortality convergence is observed due to the exhaustion of initial differences in redundancy levels. Mitohormesis has been known since the 1930s that restricting calories while maintaining adequate amounts of other nutrients can extend lifespan in laboratory animals. Recently; Michael Ristow's group has provided evidence for the theory that this effect is due to increased formation of free radicals within the mitochondria causing a secondary induction of increased antioxidant defense capacity.

Non-Biological Theories of Frailness

- **The disengagement Theory:** It gives an idea that separation of older people from active roles in society is normal and appropriate, and benefits both society and older individuals. Disengagement theory, first proposed by Cumming and Henry, has received considerable attention in gerontology, but has been much criticized.The original data on which Cumming and Henry based the theory were from a rather small sample of older adults in Kansas City, and from this select sample Cumming and Henry then took disengagement to be a universal theory. There are research data suggesting that the elderly who do become detached from society as those were initially reclusive individuals, and such disengagement is not purely a response to ageing.

- **The Activity Theory:** It implies that the more active elderly people are, the more likely they are to be satisfied with life. The view that elderly adults should maintain well-being by keeping active has had a considerable history, and since 1972, this has become to be known as activity theory. However, this theory may be just as inappropriate as disengagement for some people as the current paradigm on the psychology of ageing is that both disengagement theory and activity theory may be optimal for certain people in old age, depending on both circumstances and personality traits of the individual concerned. There are also data which query whether, as activity theory implies, greater social activity is linked with well-being in adulthood.
- **Selectivity Theory:** It mediates between activity and disengagement theory, which suggests that it may benefit older people to become more active in some aspects of their lives, more disengaged in others.
- **Continuity Theory:** The view that in ageing people are inclined to maintain, as much as they can, the same habits, personalities, and styles of life that

they have developed in earlier years. It belongs to Atchley's theory that individuals, in later life, make adaptations to enable them to gain a sense of continuity between the past and the present, and the theory implies that this sense of continuity helps to contribute to well-being in later life. Disengagement theory, activity theory and continuity theory are social theories about ageing, though all may be products of their era rather than a valid, universal theory.

- **PRISMA Model:** The PRISMA group has implemented its model in two CLSC ("Centre local de services communities) territories in the Victoriaville region (the Bois- Francs project). The purpose of this pilot project was to evaluate, using a quasi-experimental design, the implementation and impact of this model for community-living. Two cohorts of subjects in the study (ns272) and control (ns210) areas were followed and evaluated annually over a three-year period (1997 to 2000). One of the main outcomes of the study, functional decline, was defined as death, institutionalisation and significant increase in disabilities (difference of 5 points or more on the SMAF scale). In the study, there were fewer people who experienced a functional decline in the study group for those with moderate to severe disability at entry but not for the ones with mild disability. PRISMA is an innovative co-ordination type ISD model. Since it is embedded within the usual health care and social services system, this model could be more appropriate to the Canadian universal and publicly funded health care system than the fully integrated models tested so far. However, it requires a shift from the traditional institution-based approach to a client centered approach and tremendous efforts in co-ordination at all levels of the organizations.

- **Social Development Theory (Vygotsky) (1896-1934):** It argues that the social interaction precedes development; consciousness and cognition are the end product of socialization and social behavior. According to Vygotsky it is one of the foundations of constructivism. It states that the social interaction plays a fundamental role in the process of cognitive development. Vygotsky felt that the social learning precedes development. He stated that every function in the individual's cultural development appears twice: first, on the social level, and later, on the individual level; first, between people (interpsychological) and then inside the person (intrapsychological)." (Vygotsky, 1978). The More Knowledgeable Other (MKO). The MKO is normally thought of as being a teacher, coach, older adult, a younger person, and even computers. The Zone of Proximal Development (ZPD). Vygotsky focused on the connections between people and the sociocultural context in which they act and interact in shared experiences (Crawford, 1996). According to Vygotsky, humans use tools that develop from a culture, such as speech and writing, to mediate their social environments. Initially people develop these tools to serve solely as social functions, ways to communicate needs. He believed that the internalization of these tools led to higher thinking skills. Vygotsky's social development theory traditionally held a transmissionist or instructionist model in which a teacher or lecturer 'transmits' information to students. In contrast, Vygotsky's theory promotes learning contexts in which students play an active role in learning. It can be applied on elder people as well.

Psycho- Social Fabrication and Frailty Syndrome

India has become an ageing country. Yet very little national preparation has been contemplated to meet the needs of the emerging situation. The aged population in India, irrespective of caste, class, religion and region, stand out as the most vulnerable group.

Owing to the modification of the family, at both rural and urban levels, responsible and institutionalized care is not inbuilt in the social structure as it was in the past. On account of bio-psycho-socio and cultural reasons, taking care of the aged has become a serious problem. Therefore, numerous institutions which take care of the old and are managed them by government and voluntary organizations have come into being. These institutions provide a wide range of services such as residential, day, geriatric and medical care followed with recreation and counseling, amenities.

It was way back in 1948 that the question of ageing was first taken up by the United nations at initiative of Argentina, which proposed a Draft declaration on old Age Rights. The question was taken up again and placed on the agenda of the General Assembly at the initiative of Malta in 1969. But it was only in 1972 that the Economic and Social Council of the United Nations deliberated at length for the first time on the issues of ageing and the aged. However, it was only after a gap of ten years the first World Assembly on ageing could be held in Vienna in 1982 wherein the UN member-states formulated and adopted the International plan of Action on Ageing, which was later on endorsed by the UN General Assembly in same year. Again, almost a decade later, the UN adopted the principles for older persons in 1991. The eighteen principles were grouped into five clusters namely, independence, participation, care, self-fulfillment and dignity, and are best summed up in the logo, to add life to the years that have been added to life. In between the government of Malta succeeded in securing the establishment of the United Nations international Institute on Ageing (INIA) in Malta in 1987. Similarly, in 1989, the African Society of Gerontology was established. But much before that the Indian Gerontological Association was established at Jaipur in the year 1968, and the first issue of its mouthpiece, Indian Journal of Gerontology, was published the very next year (1969).

In 1990, the UN General Assembly designated October I as the International Day of older Persons for celebrating and acknowledging the contribution of older people to society. The conceptual frame work for IYOP, which is based on the International Plan of Action of Ageing (1982) and the principles for older persons (1991), was formulated and submitted by the /Secretary General to the 50th session of the General Assembly in 1995. The 1997 Operational framework also assisted in setting the scene for International Year of Older persons, 1999. The overall theme for IYOP was "Towards a Society for All Ages", which reflects a growing concern for ensuring age-integration and multi-generational relationship.

The varied and multifarious initiatives undertaken by the United Nations towards the problems of an ageing world population and the general positive attitude of the social scientists all over the world towards the specific realities of the life situations of the older persons in their societies have motivated a whole host of scholars to undertake researches on the phenomenon of ageing in its various ramifications. According to Yogesh Atal (2001) the then Regional Adviser and Principal Director of UNESCO, sets the tone of this volume in a truly international perspective, in his paper "The United Nations and Ageing" rather than focusing his attention on the aged of a specific society or nationality and/or comparing them with the aged of one or more countries he not only chronologically outlined the various international initiatives undertaken by the United Nations and its various agencies for developing an international agenda and a plan of action on ageing beginning with the deliberations of the Economic and social Council of United Nations in 1972, but also draws our attention that how West universalized the problem of the older persons, which had assumed the special significance for it as compared to the developing countries.

He further observes that since magnitude of the problem ageing and its manifestations are not the same everywhere, there cannot be a single strategy for dealing with the problem of the aged. There can only be a common commitment. However, the overall objective of the International Year of Older Persons has been a promotion of the United Nations Principles for older persons, their translation into policy, as well as practical programmers and actions. International year of older persons in 1999 prepared by the Secretary General and submitted to the 50th session of the General Assembly in 1995, which is based on the International plan of Action, adopted in 1982 and addresses the sect oral dimensions of ageing such as health and nutrition, housing and living environment, family, education and the media, social welfare, and employment/income security. He also emphasizes that research on the phenomenon of old age, and on the problems that the old encounter in different societies is still very sparse in most of the developing countries. Immediate action is needed to fill this gap so that social sciences could respond to the demands made on them by the government and other agencies dealing with the older persons.

A search of psychological and sociological abstract shows that some of the published papers in India appeared in the late 50's and early 60's. The search further showed that good bit of this research effort in psychological gerontology has come from the Department of Psychology of Sri Venkateswara University. Perhaps the first published paper on the psychological aspects of the elderly in India ws on learning in the aged (Ramamurti & Parameswaran, 1963), though there was earlier one on "Interest in music among the young and the elderly" (Ramamurti, 1956). There have been only few studies on intellectual functions of the aged in India. In an early study Rammurti and Parameswaran (1964) used reversible figures (Boring;'s Mother-in law and daughter-in-law figure and Book figure) and found a greater proportion of older men than younger men perceived the daughter in law in figures first. In learning tests too the aged were comparatively slower in learning new task. Indian studies on reaction time in elderly showed that there was an observable increase in the elderly for both simple and complex reaction time (Ramamurthy, 1982).

One of the main concerns in old age is how to be well adjusted and happy. Large number of studies in this respect has been taking place. During old age there is an anxiety and negative attitudes towards death. Narayanan (1983) found that death anxiety was related to possession of a probabilistic orientation. According to Ramamurthy (1983) individual with external locus of control had less anxiety and less negative attitude towards death than those with predominantly internal locus of control. The depression and dementia is common ailment in old age (Animasen & Ochaney, 1983; Warty, 1983). According to Venkoba Rao (1981) the depression increased physical dependency in old age. As per the study of Warty (1983) upper and middle class families in urban setting were particularly found to be victims to these pathological conditions. The defensive styles in retired persons and on-retired persons are full of depression that increases with growing age (Dixit, 1984). Menopause is also a significant even in women life in old age. The performance of the individual may decline over the increasing age. Emotional problem and insecurity increases with old age. Several Indian studies have shown the importance of attitudes of various generations towards the aged, as these attitudes greatly determine the treatment meted out to the aged in our society. Hence the attitude can be improved through social support (Jamuna & Ramamurti, 1984).

In 1990, the UN General Assembly designated October I as the International Day of older Persons for celebrating and acknowledging the contribution of older people to society. The conceptual frame work for IYOP, which is based on the International Plan of Action of Ageing (1982) and the principles for older persons (1991), was fo9rmulated and submitted by the /Secretary General to the 50th session of the General Assembly in 1995. The 1997 Operational Framework also assisted in setting the scene for International Year of Older persons, 1999. The overall theme for IYOP was "Towards a Society for All Ages", which reflects a growing concern for ensuring age-integration and multi-generational relationship.

The varied and multifarious initiatives undertaken by the United Nations towards the problems of an ageing world population and the general positive attitude of the social scientists all over the world towards the specific realities of the life situations of the older persons in their societies have motivated a whole host of scholars to undertake researches on the phenomenon of ageing in its various ramifications. According to Yogesh Atal (2001) the then Regional Adviser and Principal Director of UNESCO, sets the tone of this volume in a truly international perspective, in his paper "The United Nations and Ageing". Rather than focusing his attention on the aged of a specific society or nationality and/or comparing them with the aged of one or more countries he not only chronologically outlined the various international initiatives undertaken by the United Nations and its various agencies for developing an international agenda and a plan of action on ageing beginning with the deliberations of the Economic and social Council of United Nations in 1972, but also draws our attention that how West universalized the problem of the older persons, which had assumed the special significance for it as compared to the developing countries. He further observes that since magnitude of the problem ageing and its manifestations are not the same everywhere, there cannot be a single strategy for dealing with the problem of the aged. There can only be a common commitment. However, the overall objective of the International Year of Older Persons has been a promotion of the United Nations Principles for older persons, their translation into policy, as well as practical programmers and actions. International year of older persons in 1999 prepared by the Secretary General and submitted to the 50th session of the General Assembly in 1995, which is based on the International plan of Action, adopted in 1982 and addresses the sect oral dimensions of ageing such as health and nutrition, housing and living environment, family, education and the media, social welfare, and employment/ income security. He also emphasizes that research on the phenomenon of old age, and on the problems that the old encounter in different societies is still very sparse in most of the developing countries. Immediate action is needed to fill this gap so that social sciences could respond to the demands made on them by the government and other agencies dealing with the older persons.

In the study Gastron, Andres and Vujosevich (2001) have attempted to measure ageism transfer at school by examining answers to Help Age version of the Ageism scale in their paper "Ageism at School: images and Stereotypes of Ageing and the Old Age in Argentina". They evaluated different factors proposed in a sample of 497 elementary school teachers, of various cities of Argentina. Their findings show that more than 80 percent of the respondents have at least one prejudice against the elderly, and half of them are ageist in a middle of a high comparative degree. The more usual isolated stereotypes against the elderly were those referred to the physically visible images such as height, common diseases and damages at home. The major combined dimensions of prejudices against the aged agree with Palmore statements: older people become homogeneous as time goes by, gender differences disappeared with age, and older persons are useless, ineffective and then, unemployable. The data suggest that ageism has not only negative but positive aspects. The study confirms their prior hypotheses regarding school teachers' ageist images and then it highlights the idea that they may adversely affect the generations' attitudes about growing older.

The study of Ekstrom, Ingman and Benjamin (2001) on "Global ecology and the older adult in the 21st Century: Senior Involvement in sustainability Achievement" discuss the role that older adults throughout the world must increasingly play in contributing progress in the struggle to redress ecological balance in the world. Paradigm shifts in the thinking about senior involvement and models for senior involvement in environmental concerns has been developed by them. Drawing examples from work in the United States of America they illustrate that this pool of talented older adults is only beginning to be tapped, offering important resources for the creation and maintenance of sustainable communities.

Joseph, (2001) in his study "Influence of the old on the perspectives of the young on human life" identifies interaction of the young with the old as influencing perspectives of the former on human life. An interview questionnaire administered to 99 young persons aged under 25 in Calicut, Kerala examined their perspectives on the themes, namely meaning of human life, necessity of intimacy with land, role of material wealth in human life, meaning of God and essence of human love. The finding reveal that those who had interacted with their grandparents during childhood have a vision of human life, which is more conducive for leading a contented life than that of those who had no opportunity to interact with their grandparents. The interaction continuing beyond childhood has no perceptible influence on the perspectives of the young. The results adequately support the necessity for retaining aged parents in the family atmosphere itself and promoting their interaction with the children.

This ambiguity appears again in the contrast between the active role that senior citizens played in Roman politics, and their depiction in satirical literature of the period. Christian literature in the middle Ages also played a large part in defining the society's perception of the old, both in the image of the revered holy sage and in the total condemnation of the aged sinner. He traces the varying role of old age in public and private life from the eleventh to the sixteenth centuries. He shows that the standard of living has improved; more old people were living longer and playing an increasingly active part in the professions and society. Minois (1989) considers the interrelation of literary, religious, medical and political factors in the social fate of old men and women and their relationship with the rest of society. Further, Paoletti, (2001) in her paper "The forgotten majority: Images of older women and their lives in Italy" pointed out that in spite of older women representing the majority of older people, they are generally "invisible" in the gerontological literature. In her contribution she analyzed through set of interviews that were conducted as part of a large study in the social construction of older women's identities and revealed that such processes are connected to the women's field of action i.e. from their lives perspectives. In fact different perceptions of oneself as an older woman have great implications for one's quality of life and health. In particular the study explores some instances of institutional intervention towards the social integration of older women.

The social institution have been playing pivotal role for improving the health of the elderly people. In this respect, the family, govt. and non-government organization have playing catalytic role for restoring health of the older people. The healthy older people living with a partner feel highest quality of life and less psychological strain as compared to their counterpart those in residential homes are likely to report more frailty. Next to having a partner, older people identified family and good health as important to their quality of life and avoiding frailness. Belonging to a Church, being able to get about, membership of a lunch club and independence came next down the list, ahead of considerations such as companionship or security.

Over the whole range of socio-psychological factors directly affecting people's perception of their quality of life, the social environment - including home, safety, finances, services, leisure, environment and transport - is the single most important. Those in residential homes gave the lowest ratings on all counts when compared with anyone living with a partner, family and friends. The study says that physical disability was seen by them as a factor in a lower quality of life, being associated with lower mental ability. The more disabled were likely to be in sheltered or residential accommodation, and were more likely to be depressed or have lower self-esteem. The most important factors determining the older people's sense of quality of life, besides the social environment, were how healthy they felt (although all were actually in good health, freedom from depression, good mental faculties and personal optimism. Depression further influences quality of life by affecting mental alertness, and worries about health were the main factor affecting depression and social life.

The Continuity refers to the organized, co-ordinate and steady passage of individuals through various elements in a system of care and services. It comprises of two aspects: the short-term aspect (synchronic) that relates to the application of an intervention and concerted and co-ordinate service plan over a given period. The long-term aspect (diachronic) relates to monitoring and harmonizing intervention and service plans over a protracted period. This later aspect has also been called "longitudinality" by Starfield. Integrated Service Delivery (ISD) programmes have been developed to improve continuity and increase the efficacy and efficiency of services, especially for older and disabled populations.

CHANGING FAMILY-COMMUNITY NETWORK AND ELDERLY FRAILNESS

Adam was young, fair, and joyful;
But having spurned the commandment,
He became unhappy, old, and fading,
Bearing the weight of years
And a load of miseries

Minois, 1989

Dynamics of Ageing

Like Adam everybody in the universe glow, progress and perform better in their youth and lost everything in the old age. "Old age", the very word brings to mind white hair, stooped shoulders, an uncertain gait, and certain isolation. No one knows when old age begins, and decline. The process of ageing varies with individuals; some begins to look old at 59 years, while other looks young even at 65 or 70 years, therefore is a universal process. Human life from conception to death is a complete sequence of events. Human concern about the phenomenon of ageing is age old. Man has, for long, been trying to unravel the mystery of growth, ageing, and death and perhaps the after life. It indeed is ironical that though India has an aged population (55+) larger than that of the United States or any other Western country, documented facts of the psychological status of elderly here is extremely meager. It is only the recent past that there has been a semblance of an effort by Psychologists and sociologists to delve deep into the status of aged.

For convenience of description the whole life cycle has been divided into four distinctive but successive stages. They are juvenile age, young age, middle age and old age. These four main stages in human life have been compared with four main reasons of the year i.e. spring, summer, autumn and winter. Spring and summer represent the energetic and joyful childhood, warmth and vigorous youth respectively. Autumn denotes the middle age of maturity and ripeness. After this comes the winter of human life which is the old age when one's physical strength begins leaving him and youthful vitality starts yielding to a sense of despair (Sinha, 1989). The teeth start falling, the hair turn grey, skin becomes wrinkled, vision becomes poorer, ears need hearing aid and there is decrease in muscle strength. This is the time when many a man becomes three legged (the third leg bring the stick). It is from here that the problems of old people start arising. Most of the people anticipate old age with trepidation will it be the "golden year" or "statuary servility"? There is a special and not realistic fear, by young and old, that many of the last years of life will be spent in ill health, with some chronic diseases that will limit activity and possibly even impair reasoning abilities. The elderly are a precious asset for any country with rich experience and wisdom; they contribute their might for the sustenance & progress of the nation. We must today think of sound old age in society, which has to have a positive attitude towards the aged.

The peak of human development lies in the attainment of social maturity. The ability to reciprocally function in a complex social milieu is not only an indication of social maturity; but, it is also associated with happy disposition. It is an important aspect of adjustment at all stages of human development especially in the advanced stages. Among elderly people with diminished roles, interactions in their peer and social groups it is likely to reduce the feelings of isolation or rejection, satisfy some of their emotional needs and enhance their morale. There are two contrasting theories of successful ageing. The better known is the *"Engagement theory"* which states that continuation of activities and attitudes of middle years into old age is conducive to happiness and successful aging (Havighurst, 1961). The other theory is the Disengagement theory as proposed by Cumming and Henery (1961) which states that individual prefers to give up his earlier activities and social roles and society tends to take away his earlier professional and social responsibilities both not minding but mutually and willingly seeking the disengagement. Maddox (1964) emphasizes on psychological and social withdrwal model and states thata disengagement process is intrinsic and inevitable is disputed and there are factor such as sex, health, profession, intelligence, personality type and life style which modify the disengagement process. According to Seal (1979) elderly people suffer with national, special and personal problem. In the word of Rao (1979) the aged people suffer with socio-economic condition like poverty, breaking up of joint family system and poor service for the aged pose a psychiatric threat to them.

The disengagement process is necessary for successful ageing. Paintal (1976) further explored engagement theory and successful ageing on 94 Hindus aged 45 year to 74 year. Result revealed that the circle of friends becomes narrower in old age but well adjusted men have more friends, frequencies of meeting with friends after 45 year however well adjusted men do not show decline in it. Adjusted person attend social gathering as compared to their counterpart (Paintal, 1979). But there are certain factors as pointed out by those depict successful ageing. More specifically it deals with physical and social relationships, leisure and recreation, economic security and religion for better adjustment. The study was conducted between 45-74 years in Bangalore; wherein most of the elderly people found adjustment problem. Good physical health was related to better adjustment problem as compared to poor health. Participation in leisure activities also showed better health and adjustment whereas economic security did not discriminate the two adjustment groups. Paintal (1978) in another study reported better correlation between life satisfaction index and Chicago attitude inventory while assessing elderly. In the words of Ghosal (1982) the problems of old age tends to be multiple and complex rather than single and simple involves social factors those play important role in their life. They perceive psychosocial changes in the older individual's life span such as loneliness, isolation, existential sadness and so forth are dispread enough. They suffer from economic security, housing, health services, transportation and facilities for recreation (Narain, 1979)

One of the main concerns in old age is how to be well adjusted and happy. It is a documented factor that consequent on the attainment of old age, the individual experiences changes in the roles and statuses, both in the home life as well as in occupational and social life. The happiness and satisfaction would to a large extent, depend upon how successfully he meets this challenge. Studying adjustment trends across theh middle years and old age (Ramamurti, 1968a, 1971a) found that among the urban aged men there was deterioration in adjustment with increasing age. The trend indicated that among those who were employed, the period just prior to retirement showed an increase in maladjustment. However, there was a gradual improvement in adjustment in post-retirement period. A study of rural women conducted by Jamuna (1984a) and Menachery (984) also found a similar deterioration of adjustment with age in older year. Further a study of *Ramamurti (1970) observed that life satisfaction declined significantly around retirement period.* A large number of studies have been conducted on adjustment (Ramamurti, 1968a, Siddaiah, 1970; Desai & Naik, 1971; Paintal, 1970; Shalini Bhogle et al, 1978, Anantharaman, 1976, Chandrika & Anantharaman, 1982; Hrudayanatha & Reddy, 1984; & Menachery, 1985). Beside this other variable in old age are also causing rigidity and inflexibility (Ramamurthi, 1970), husband wife communication (Jamuna & Ramamurti, 1984), marital satisfaction (Prakash,s 1983), attitude toward future and death (Anantharaman, 1980), role activity (Jamuna, 1984), family type and socioeconomic status (Desai & Naik, 1970) are related to better adjustment.

Indian research on psychosocial aspect of aging is s till in its infancy. Much of the work that has been carried out so far has centered on the study of problem of adjustment in old age and on attitudes towards the aged. Studies on cognitive functions, learning and personality are comparatively less. In the last ten years there has been a step-up of research work on the psycho-social aspects of aging. For improving psycho-social conditions of the aged people counseling, anticipatory preparation, use of media in the enhancement of social and families values towards a better stat us for the aged, and the development of a data base on the Indian aged to support research and welfare activities of the aged can prove beneficial (Jamuna &

Ramamurti, 1984). Neglect of the aged, earlier considered a western aberration has now spread to Asian countries with the break of joint family system in urban middle class (Probe India, 1982). The aged feel neglected in their own house where they once yielded power in not too distant past.

The low proportion of the old in the population of the developing countries notwithstanding the actual size of the older population is already 61 percent of the world's older persons population, and the major increase is expected to occur in the future in the less developed countries. *By the year 2025, more than 70 percent of older persons are projected to live in the developing world.* It must, however, be noted for record that the question of ageing was first taken up by the United Nations way back in 1948, at the initiative of Argentina, which proposed a Draft Declaration on Old Age Rights. Again, in 1969, the question was taken up and placed on the Agenda of the general Assembly at the initiative of Malta. India responded only to the questionnaire for the second appraisal held in 1984.

Some of the old person are quite self-efficacious maintain their physical, socio-cultural and psychological wellbeing whereas the mass other living especially in the rural and urban areas lost their balance and become vulnerable to psychopathologies. As one becomes weak from mind as well as from body the nearer and dearer one also wants to forget them and even expect their early death. Since the elderly are very rich from their cognitive processes understand the attitude of their younger as well as surrounding societies. Since time immemorial the society including academician, policy makers and politician are not bothered about the problem faced by the old people. In this respect they are suffering from time immemorial.

Beginning with the ancient Egyptian and biblical sources, Minois (1989) traces the changing conceptions of the nature, value and burden of the old. He shows how, in ancient Greece, the cult of youth and beauty, on the one hand, and the reverence for the figure of the Homeric sage, on the other, created an ambivalent attitude towards the aged. Moving the older population from a sedentary to an active lifestyle is a challenge for health and mental health practitioners. *Life expectancies for men and women are increasing (U.S. Bureau of the Census, 1997), and the over-65years of age group is projected to double in size by 2020* (Bokovoy & Blair, 1994). *Within the population of older Americans, the number of individuals over 85 years is expected to quadruple* (Zedlewski, Barnes, Burt, McBride, & Meyer, 1990). From a social perspective, *individuals in the 8th and 9th decades of life are expected to be involved in the health care system to an unprecedented degree if population growth continues as projected* (U.S. Bureau of the Census, 1992). *According to Elango (1998) there are more elderly males as compared to females those suffer from health problems.* In both developed and developing countries more women than men are widowed. In contrast, there are no countries in which more than 40% of men aged 75 years and older are widowed (WHO, 1995).

According to Emile Durkheim (1912,1965), ideas about the ultimate meaning of life and ritual ceremonies that express these ideas arise out of the collective experience of individuals (Kart, 1995). In the words of Ray (1975) 80.1% of aged performs Puja and prayers during their leisure time. Further Gurumurthy (1998) observed in a study of the rural area of Karnataka that 66% of the aged owned some or the other property. Out of them .2% owned a house, 59.9% both house and agricultural land and 2.2% own all three types of property such as house, agriculture land and plot. Now there are inter generational gap in Indian society in respect to psychological manifestations (Sinha, 1972).

In the study, Rook etal (1990) interviewed 162 community residing older adult and assessed social control, health risk taking, psychological functioning and interpersonal satisfaction as the main factors behind their suffering. Similarly Murphy (1982) found onset of depression in elderly associated with major life events. Earlier Cameron (1945) noticed certain clinic-psycho-pathological trends in geriatric population. Similarly Muller-Spahn, etal.1994) observed that depressive syndrome and dementia are the most frequent psychiatric disorders in elderly people. Chen (1994) found that the prevalence of hearing impairment increases with age. As per Heikkinen et al (1995) there are recent stressful life event behind their suffering. Similarly Senstock, 1993) found victimization in elderly that includes psychological abuse, neglect and exploitation as well as more life threatening physical abuse and neglect.

Venkoba Rao, (2001) in his article "Euthanasia and Psychiatry" has discussed euthanasia with reference to issues relating to law prevailing in different countries. Psychiatric aspects of terminally ill patients wherein euthanasia is considered are detailed. Depression in terminally ill may cause suicidal ideation and this needs to be recognized and treated. The two options for the terminally ill "hospice versus euthanasia" were described. A brief reference is also made to psycho spiritual aspects of Indian attitude towards death and dying. The deaths of Bhishma and King Parikshit are referred to illustrate this. Bhatnagar, in his paper "Death, Dying and Living" observes that normally death is considered to be an evil and something to be feared. However, there is a strong and prestigious strand of thinking which derobes death of its fear and ferocity and presses it into the service of living raising its quality. With the related ideas a distinction is introduced in the present essay between "dying of" and dying for", showing that the later is essential for the higher quality of life. Besides this there are certain factors those are biological, social, psychological and spiritual, in nature as depicted below.

Bio-Psycho-Socio-Cultural Stressors and Elderly Well-Being

Aging is a normal inevitable and universal phenomenon. Commonly speaking; it means the various effects or manifestation of old age. While they have been usually perceived as biological, there is also deterioration in mental capabilities and social adaptability. Ageing has thus three aspects, viz., and biological, physiological and social. The word ageing thus implies a wide variety of processes, these include the biological inevitability of growing old, the psychological reality of growing old, and the social content, which provides a wealth of meanings in which growing old is understood and made sense of. Some aspects of this process are inevitable.

There is a disagreement about how to define old. Should the aged be defined chronologically (how many years one has lived) or functionally (what one is capable of doing in a specific area of functioning) or socially (one's role activities in society). Generally, chronologically definition is taken arbitrarily as it is more convenient and simpler. According to Malher, Director General World Health Organization (1982), "Ageing" is not simply a physical process but a state of mind and today we witnessing the beginning of a revolutionary change in that state of mind. For biologist and medical scientists ageing refers to deterioration in psychological capabilities. According to Cowdry (1942) the changes found in aged persons as structural alteration due to infections, toxins, traumas and nutritional disturbances or inadequacies give rise to what are called degenerative changes and impairments. Challam (1982) has stated that generally changes take place in four main spheres-physical, behavioural biological and intellectual. Empirical studies have indicated that definitions relating the ascription of old age as well as the characteristics associated with the aaged vary in different social categories.

According to Ahammier and Baltes (1972), Neugeston and Modre (1968) "Biological ageing is most prominent and has figured widely in common expression and even in scientific literature with advancement in chronological age, an individual passes through different stage of life cycle as described above. Every one attains old hood at a particular age. Nair (1982) has reported that ageing varies from individual to individual One may lapse old at 50 or still young even as an octogenarian. It is here that physiologists and social scientists have their role to play. One may live a happy youthful life even at an advanced chronological age by developing appropriate attitudes and styles of life. Growing old is an inevitable phenomenon and nobody can stop the wrinkles, graying hair and the gnarled hands? Tuckman & Lorge (1956) mentioned that both young and old people look upon old age as a stage characterized by economic insecurity, poor health, loneliness, resistance to change and failing physical as well as mental powers. Ramachandaran (1981) has found that the family and living conditions are significant factors affecting the mental health of the elderly subjects.

The third important aspect of ageing, as mentioned earlier is psychological to which the present research is mainly devoted. Old age has many psychological manifestations.

Mental disorders are common in the United States and internationally. An estimated 26.2 percent of Americans ages 18 and older - about one in four adults - suffer from a diagnosable mental disorder in a given year (Kessler, Chiu, Demler, & Walters, 2005). When applied to the 2004 U.S. Census residential population estimate for ages 18 and older, this figure translates to 57.7 million people (U.S. Census Bureau Population Estimates by Demographic Characteristics, 2000). Even though mental disorders are widespread in the population, the main burden of illness is concentrated in a much smaller proportion - about 6 percent, or 1 in 17 - who suffer from a serious mental illness 9 Kessler, Chiu, Demler, & Walters, 2005). In addition, mental disorders are the leading cause of disability in the U.S. and Canada. Many people suffer from more than one mental disorder at a given time. Nearly half (45 percent) of those with any mental disorder meet criteria for 2 or more disorders, with severity strongly related to co morbidity (Kessler, Chiu, Demler, & Walters, 2005).

It is expected that at least 40 percent of the population over 75 will need extensive health care in general and mental health care services particular in their lives. The public has a negative view of nursing home placement that has, to some extent, been confirmed by research finding that the health of a frail older person deteriorates each time he or she is moved. The Aging in Place model of care for the elderly offers care coordination (case management) and health care services to older adults so they will not have to move from one level of care delivery to another as their health care needs increase. University Nurses Senior Care (UNSC) is the service entity of this project and provides as its core service care coordination with a variety of service options. These options include care packages or services at an hourly rate to meet individual client needs. The Aging in Place project will be evaluated by comparing project clients to residents of similar acuity in nursing homes and to similar clients receiving standard community support services. Data from this project will be important to consumers, researchers, providers, insurers, and policy makers (Myers & Hwang, 2000).

To examine the association of psychosocial adaptation status with vision-specific health-related quality of life (HRQOL) and the role of psychosocial adaptation in the linkage between visual impairment and vision-specific health related quality of life outcomes among older adults with visual disorders. Design and methods in this cross-sectional study, older urban adults with visual problems (N = 167) were interviewed using a structured questionnaire to assess their self-reported visual function, general health, psychosocial adaptation status, and vision-specific health related quality of life. Performance-based measure of visual function marked by distance visual acuity was clinically conducted by ophthalmologists. Results revealed that psychosocial adaptation status was significantly associated with vision-specific Health related quality of life including the domains of mental health symptoms due to vision and dependency on others due to vision. The results also showed that psychosocial adaptation status could buffer the effect of visual impairment on vision-specific HRQOL, including the domains of social function, mental health, and dependency. Psychosocial adaptation status was significantly associated with multiple domains of vision-specific HRQOL. The findings have significant implications for health education and psychosocial intervention for older adults with age-related vision loss (Wang, & Chan, 2009).

But living longer is not enough - it is the quality of that life which counts. What emerge from the research are core components in quality of life. Control and mastery are shown to be extremely important," life among older people is about much more than just health and physical functioning. Quality of life needs of older people are not really different from those of other age groups - we all place value on health, socio-economic status, social networks, satisfaction of emotional needs and other key factors. Environmental factors - inside and outside the home - and leisure and mobility are also important in determining quality of life in older people, but there has been far less research in these areas.

In a study, Reisine and Locker (1995) societal indicators such as work loss due to dental problems to describe the social impact of oral disease. A limitation of this method is, while useful for indicating trends in uptake of health care services, societal indicators give little information on an individual level. Similarly Dixon (2001) in her paper "Cross-Culturally Comparative Structural Constraints Affecting the Social Aspects of Aging" observes that aging is a universal process beginning at birth and differentially constrained by the socio-cultural location. This paper proposes a cross-culturally valid theory of life cycle involvement in family, education, work, and leisure. The underlying hypothesis is that the web of life cycle involvement has impact on social psychological factors like life satisfaction, self-esteem, and the quality of the aging process. Much theories have been based on the Western European middle class "average man", whose successful "aging" is accomplished by timely "disengagement" or continued involvement in "activities". Although the aging literature is replete with studies of social problems, using age, social class, ethnicity and gender as independent variables, much of that variation could be explained by the structure of the people's life cycle involvement.

In their paper Samuelsson and Johnson (2001) "Centenarians in France, Georgia, Hungary and Sweden: social characteristics" state that social living conditions have been found the great importance for longevity. In this paper earlier and present living condition were compared across the four samples. Centenarians in Georgia and Hungary had more social network resources than centenarians in Sweden and France. The Centenaries seem in general not to be isolated and only about 10% reported "often" feeling of loneliness. Public support dominated in Swedish group and to some extent in the French group, while informal support dominated in the Hungarian and Georgian sample. The results are partly related to cultural and macro-social dimension. Taken together, the differences among the four samples seem to be more than the similarities. The possibility for making distinct conclusions about the importance of social factors for longevity is thus fairly weak.

In further study de Oca (2001) in her paper "Discourses, voices and Visions on Aging in Mexico City" describes about the actual meaning of to be in Mexico older men. This meaning is not an isolated phenomenon. She thinks that this is a consequence of the discourses constructed about the aged and the demographic, political and economic processes. The paper has two sections. In the first, she analyses the contemporary Mexican State regarding the population policies and ageing process. In the second, she analyses the contradictions between the institutional discourse and the public policies. The social perception on the aged people appears as an effect of these elements at the macro level. This social construction on the aged produces a vulnerable condition in the older persons. At the economic and cultural levels exists a conciliation that justifies a new form of discrimination to the aged and the older people.

Markus (2001) in her paper "Uniqueness of the Israeli over 70 Age Group" demonstrated that the variety of cultural, ethnic and educational backgrounds of the older Israeli society and how their different needs are being met. Various frameworks and types of support services that are offered are described. It also includes the wide range of "homes" for the elderly that are available. The paper also considers the changing needs of this society as aspects of gender and history will change their interests, demands and needs of the next generation of elderly and the generations to follow. According to Hossain (2001), in his paper there are certain has studies some socio-demographic characteristics involved in of the elderly persons well-being. He observes that the gradual advancement of industrialization, modernization and urbanization in Bangladesh has caused a radical impact on traditional family life and traditional care system of the elderly persons. Demographic characteristics, socio-economic characteristics, morbidity statistics and living arrangements of the elderly are also incorporated in the direction.

Marin, in her paper "Successful Ageing-Dependent on Cultural and Social Capital?: Reflections from Finland" opines that the increase of elderly population has aroused discussion and debates on several age-and generation-related issues, such as political, social cultural and medical ones. The idea behind this article is to tackle with changing position of elderly people in society, especially concerning their power and autonomy. It also deals with the cultural and social resources the elderly people have in the modern society. The viewpoint chosen is arising from Pierre Bourdieu's theories which deal with different kinds of resources (capital) that influence people's position in power structures and social life. These theories are mostly neglecting the age question, as well as the question of gender (Marin, 2001).

Bhawsar, (2001) in his paper "Population Ageing in India: demographic and Health dimensions" presents demographic and health dimensions of ageing population (60-plus) in India and its major states. Various indicators such as changes in age structure, sex ratio, rural-urban residence, marital status, support ratios and causes of deaths among the elderly were analyzed for different census years and discussed in detail. It shows that the process of population ageing has already started in India. With the levels, the size of the younger population in shrinking leading to the ageing of the population. In some of the Indian states (Kerala, Goa, Tamil Nadu), the pace of increase is much faster while in others (Assam, Bihar, U.P) it is moderate. Trend in sex ratios and greater life expectancy of females indicate that feminization of elderly population of non-working age relative to the working age in all the states. It is also foreseen that due to rapid population growth and ageing, more proportion of elderly will be concentrated in near future, particularly in North Indian states. Hence the changing life style and modernity as well as numerous of her are exerting pressure on human well-being by increasing all static load that in turn are associated with their psychological frailness.

Changing Family-community Network and Elderly Frailness

Swiss scholars Vollenwyder, Bickel, d'Epinay and Maystre (2001) in their paper "changing Family Networks and Relationships in Two Swiss cohorts of elderly (1979-1994)" have analyzed the changes occurring in the structure of the family network of older people and their family interactions over a 15-year period. Data are from two Swiss cross-sectional studies carried out on a random sample of people aged 65-94 years, in 1979 (N=1,519) and 1994 (N=1,583). Both were conducted in two specific regions namely the alpine canton of Valais, steeped in rural tradition, and the metropolitan area of Geneva. Their results show a widening of the family circle due to increased life expectancy and a sharp drop in the number of childless families in the urban region because of the baby boom; at the same time, family ties multiplied contrary to the widely held view that family relationships are weakening. Furthermore, while in 1979 each region had its own specific family culture, an alignment of family structures and relationships in the two regions took place over the 15-year period. Finally, an analysis of the instrumental support relationships based on the 1994 survey reveals the strong involvement of the elderly in their families.

Sivakumar, (2001) Integration of Strategies for the care of older persons: a Cross-Cultural Study" observes that aging of population as a branch of socio-psychological and social work study has been getting worldwide attention. The purpose of the study is to link the strategies of social development and the issues concerned with aging of population. Integration of strategies is essential for the comprehensive and sustained social development of older persons. A multidisciplinary team approach consisting of the geriatrician, the demographer, the gerontologist, the professional social worker, the clinical psychologist, the social work journalist, and geriatric nurse in well organized geriatric centers and their community outreach work through popular participation is a proposition for the future. The role of the family as well as health care for a disability free life expectancy is essential. The aged population (60+) in Kerala was 9.2 percent in 1992 93 as per National family Health survey. In china, it was 9.4 percent in 1995. The proportion of aged population in Kerala (India) and China are almost similar. It is found that among older persons a large percentage needs medical care for variety of illness. Psycho-social work counseling, continuum of care programmes, and medication are essential Sivakumar, (2001).

A study of Jamuna, (2001) in an "Intergenerational Issues in Elder Care" observes that like many eastern cultures, in India too, family is the primary care provider of the elderly. The family consists of members of different generations who come under the influence of ties they live in during their formative years. More often than not the interests and values of these persons may vary and not infrequently result in conflict. Therefore, amity between the generations is an essential requirement for good care of the elderly. This paper discusses intergenerational perceptions of elder care and nuances of the care giver-care receiver relationships based on empirical studies.

Analysis of Prasad and Bould, (2001) in their discourse "Intergenerational Households of Indian-Americans" observe that the Indian-American family experiencing dramatic changes in intergenerational experiences which have transformed all roles, but specially the role of adult daughter or daughter-in-law and mother or mother-in-law. With the opening up of immigration in 1965 to professional workers, as well as the educational opportunities in the universities in the U.S. India experienced a "brain-drain". The number of young Indian professionals emigrating to the U.S increased considerably especially with the expanding opportunities in 70's. Now, the parents of these immigrants are making decisions about where to settle in their retirement years. Many have come to live with their children in U. S. This paper analyses the experience of the relations between aging parents and their adult children of this diasporas. Further, study of Rani (2001) on "Institutional Care of the Aged" notes that the care of the aged traditionally has been the concern and responsibility of the family. In fact, the aged were considered to be a blessing to the family. But, in the present scenario, new factors have emerged which have weakened the tradition. The results reveal that aging is not just biological but psychological and sociological too.

Chen and Eva Lu (2001) in their paper "Community Care and Social Support for Older persons in UK, USA and China: An international perspective" observe that community care and social support are two leading approaches to elderly welfare. Their paper reviews the experiences of different nations with the purpose to illuminate the cruxes and identify potential breaches for making headway in the next century. The UK case shows that community cafe is hardly geared toward the needs of older persons if care in the community simply means care by the community without care for the community. The US case demonstrates that social support study will help clarify community care mechanism, though how to enhance it through intervention is far from settled. The Chinese case also indicates the need for formal support including institutional care. Community care, however, tends to be the first choice if social support networks are strong and well maintained. The strategy of "networking" points to a possible breakthrough in improving social support for older persons in all countries. Similarly, Betty Havens and Madelyn Hall, in their paper "Social Isolation, Loneliness, and the Health of Older Adults in Manitoba, Canada" notice that social isolation and social loneliness are often considered to be problems of growing older. Their study uses data from the Aging in Manitoba study to explore the association of isolation and loneliness with health among a population of much older Canadians. Levels of loneliness and isolation among this sample are high. The findings show that those who perceive their health as poor, have more than four chronic illnesses, and who live alone, are cumulatively more than five times as likely to express high levels of social loneliness. Those who have very low levels of social contact tend to be female and older, regardless of health status.

In the same time, Swiss scholars Vollenwyder, Bickel, d'Epinay and Maystre (2001) in their paper "changing Family Networks and Relationships in Two Swiss cohorts of elderly (1979-1994)" analyzed the changes occurring in the structure of the family network of older people and their family interactions over a 15-year period. Data are from two Swiss cross-sectional studies carried out on a random sample of people aged 65-94 years, in 1979 (N=1,519) and 1994 (N=1,583). Both were conducted in two specific regions namely the alpine canton of Valais, steeped in rural tradition, and the metropolitan area of Geneva. Their results showed a widening of the family circle due to increased life expectancy and a sharp drop in the number of childless families in the urban region because of the baby boom; at the same time, family ties multiplied contrary to the widely held view that family relationships are weakening. Furthermore, while in 1979 each region had its own specific family culture, an alignment of family structures and relationships in the two regions took place over the 15-year period. Finally, an analysis of the instrumental support relationships based on the 1994 survey reveals the strong involvement of the elderly in their families.

A study of Maria-Teresa Bazo, on "Family and Community Care for elderly people in Spain" deals with family health care policies. It is embedded within the change processes that are currently occurring. These processes are related to demographic, family, economic and social change, and affect the European countries as well as other developed societies. Currently, a common concern in these societies is how to deal with the changes that needs to be made within the welfare systems, in order to improve both economic and social development. Following such analysis, some research results and conclusions have been presented. This research was carried out by the author in the Basque country, Catalonia and Madrid, using qualitative methodology. She feels that this topic, namely, family health care for frail elderly persons, has not been sufficiently studies in Spain. The results highlight the loneliness and, some cases, social isolation of women carers in Spain. The vast majority is caring without neither help of formal social services nor voluntary services. The main support they receive, both instrumental and expressive, is provided by their husbands and i some way by their husbands and in some way by their young and adult children. Sometimes they experience contradictory feelings: the consequences of the caring may result in a satisfactory experience, but it may be a traumatic one too. The principal conclusion puts down the family as the first welfare system for everybody, in spite of the large economic, social and cultural changes that have taken place in Spain in the last twenty years. Another finding refers to the social construction of the care role based on social expectation on women as being the natural careers, as it has been found in other societies (Bazo, 2001).

This article argues that the time is right for nurses in the UK to become the case managers in all healthcare settings. The re-launch of family health nursing, as a model for the organization and delivery of nursing care in the community, and the advent of the GP practice-based self-managed integrated nursing teams, offer the means by which to take up the opportunities presented by recent legislation and the national strategies for promoting partnership working and collaborative practice. Nurses could approach this by combining their current involvement with developing the single assessment process for older people with the overall development of inter-professional collaborative practice across all boundaries in health and social services. Despite the new opportunities, this will not be straightforward because of the still existing problems associated with the health and social care divide. In order to generate high quality care, it is imperative for nurses and their patients that the profession gains control and ownership of its own policy, remit and practice. Nursing care should be defined according to the patient's condition, so that their dependency level, diagnostic picture and potential for rehabilitation govern the eligibility criteria for health or social care and not the level of technicality in the task itself (Kesby, 2002).

Sometime the old people are highly involved in the night dream that threatened them. Differences across the lifespan in allusions to the body are the focus of this review of studies examining ideal and nightmare (or feared) self-descriptions. Among adolescents, desires for physical beauty and physical metamorphosis (e.g., maturation into a shapely adult body) are at their zenith. Young people also frequently mention ideals involving body as commodity such as desires to barter physical might or appearance for money as a professional athlete, entertainer, or fashion model. In nightmare self-descriptions, the body is depicted as a siren among the young: fears of falling victim to bodily addictions and carnal cravings crest around young adulthood. In contrast, other body themes such as fears of physical and mental incapacitation become more prevalent with age. Desires for bodily strength and physical health continue in force through middle adulthood. Main effects of sex and ethnicity are also reviewed (Bybee & Wells, 2006). But the preventive measure and positive and positive attitude towards it may prove beneficial for restoring their well-being.

Preventive Measures and Elderly Well-Being

The problem of ageing is a global problem in the sense that it is experienced by all the societies, but its magnitude and its manifestation are not the same everywhere. What is common is the fact that extended longevity due to decline in death rate and falling fertility rate is the universal cause of rise of the ageing population. This rise may affects the demography of the labour market, patterns of consumption and production, trends of savings and investment, priorities in public spending, and delivery of social services. But the similarity ends there. Just how each of these areas will be affected will differ from society to society. As such, there cannot be a single strategy for dealing with the problem of the aged. There can only be a common commitment.

The UN principles for older persons were evolved in recognition of the fact that "in all countries, individuals are reaching an advanced age in greater numbers and in better health than ever before". They emphasized the need "for a variety of policy responses because of tremendous diversity in the situation of older persons not only between countries but within countries and between individuals". They also asserted that "scientific research disproves many stereotypes about inevitable and irreversible declines with age", and recommended that "opportunities must be provided for willing and capable older persons to participate in, and contribute to, the ongoing activities of society". The eighteen principles enunciated by the United Nations fall into five clusters, namely, independence, participation, care, self-fulfillment, and dignity.

The phenomenon of old age is going to assume increasing importance with the changes in the age pyramid due to rising life expectancy rates. Policies will be needed not only for those special groups among the old, such as the poor, but for the entire older age group so that they do not suffer from poverty, elder abuse, humiliation, ill-health and the like, and are facilitated to lead a life of dignity. This would require additional economic resources, and strategies to overcome the existing impediments to sharing. The redistribution of resources to accommodate the needs of the elderly will also affect other age groups, particularly the children. Therefore, the strategies devised for the elderly should not be seen in isolation, but as part of an overall strategy for social development, and with the involvement of all partners-the government, the civil society and NGOs, the scholarly community, particularly social scientists, and the international inter-governmental organizations. The preventive actions targeting at community-dwelling frail older people will be increasingly important with the growing number of very old and thereby also frail older people. The U.S. Department of Health and Human Services (1996b) reports that "as many have been attributed to a lack of regular physical activity" (p. 24).

Further the empirical literature on recent studies of case/care management interventions for community-dwelling frail older people and especially with regard to the content of the interventions and the nurse's role and outcome of it. Very few of the interventions took either a preventive or a rehabilitative approach using psycho-educative interventions focusing, for instance, on self-care activities, risk prevention, health complaints management or how to preserve or strengthen social activities, community involvement and functional ability. Moreover, it was striking that very few included a family-oriented approach also including support and education for informal caregivers. Thus it seems that the content of case/care management needs to be expanded and more influenced by a autogenic health care perspective. Targeting frail older people seemed to benefit from a standardized two-stage strategy for inclusion and for planning the interventions.

A comprehensive geriatric assessment seemed useful as a base. Nurses, preferably trained in gerontological practice, have a key role in case/care management for frail older people. This approach calls for developing the content of case/care management so that it involves a more salutogenic, rehabilitative and family-oriented approach. To this end it may be useful for nurses to strengthen their psychosocial skills or develop close collaboration with social workers. The outcome measures examined in this study represented one of three perspectives: the consumer's perspective, the perspective of health care consumption or the recipient's health and functional ability. Perhaps effects would be expected in all three areas and thus these should be included in evaluative studies in addition to measures for family and/or informal caregiver's strain and satisfaction (Hallberg, Kristensson, 2004).

The aim of Moyle, Clarke, Gracia, Reed, Cook, Klein, Marais and Richardson (2010) study was to explore the experience and strategies of mental health well-being through resilience in older people was across the four participating countries. The findings highlight the importance of maintaining mental health well-being through resilience. Although there were some variations between countries, these strategies for maintaining well-being transcended culture and nation. An appreciative inquiry approach was used. A convenience sample of 58 people over the age of 65 years from Australia, UK, Germany, and South Africa were interviewed. Data were analyzed using thematic analysis. Results revealed that the participants described their experiences of mental health well-being in relation to: social isolation and loneliness; social worth; self-determination; and security. Strategies utilized include promoting resilience by maintaining community connections and relationships, keeping active, and emotional, practical and spiritual coping. Psychologists have examined coping skills in the elderly. Various factors, such as social support, religion and spirituality, active engagement with life and having an internal locus of control have been proposed as being beneficial in helping people to cope with stressful life events in later life. Generally the relationship between quality of care and quality of life of frail older persons who are dependent on external support and care revealed that the quality of life in old age has become a key issue within gerontology and health and social care research (Vaarama, Pieper, & Sixsmith, 2004).

Given the complex needs of frail older people and the multiplicity of care providers and services, care for this clientele lacks continuity. Integrated service delivery (ISD) systems have been developed to improve continuity and increase the efficacy and efficiency of services. The Program of Research to Integrate Services for the Maintenance of Autonomy (PRISMA) is an innovative ISD model based on coordination. It includes coordination between decision makers and managers of different organizations and services; a single entry point; a case-management process; individualized service plans; a single assessment instrument based on clients' functional autonomy, coupled with a case-mix classification system; and a computerized clinical chart for communicating between institutions and professionals for client monitoring. Preliminary results on the efficacy of this model showed a decreased incidence of functional decline, a decreased burden for caregivers, and a smaller proportion of older people wishing to enter institutions (Hébert Durand, Dubuc, Tourigny, 2003 or PRISMA Group, 2003)

A study of de Castro and Castro (2001) on Coping in Old Age: Considerations from the point of view of case Studies from Brazil proposed to understand coping mechanisms as a subjective disposition to self-differentiate. As such, coping mechanisms at old age are held to be the result of a lifelong process which enables old people to better adjust themselves to the multiple challenges of the ageing process. Two case studies in the form of life-histories are presented and analyzed in order to illustrate how throughout one's lifetime inner resources and potentialities can be restrained or expanded. Finally, it is argued that cultural dispositions can legitimate a view of the self as multiple so that individuals can fare better in managing the challenges that old age imposes.

Further Kate Davidson, (2001) in her paper "Reconstructing Life after a Death: psychological adaptation and Social role Transition in the Medium and Long Term for Older widowed Men and Women in the UK" looks at how older men and women in the United Kingdom reconstituted their lives after widowhood in the medium and long term. It examines issues of coping strategies for successful independent living for this generation, some of whom are alone for the first time in their lives. The paper looks only fleetingly at the experiences of widowhood in the first months- its purpose being to set the scene for how widows and widowers describe their personal development over time. The paper explores the differences, both inter-and intra-gender, in travelling what author has termed the ontological coherence-ontological chaos continuum. The paper also explores the meanings of the relationship between solitary living (aloneness) and feeling desolate (loneliness) and how these might be experienced differently by older widows and widowers.

The last three decades have produced a number of studies that have examined the antecedents and consequences of self-efficacy in elderly people. In general, reviews of the literature describe self efficacy as an individual's judgment of his or her ability to successfully complete a chosen task (e.g., Bandura, 1977, 1978, 1997; Gist, 1987; Gist & Mitchell, 1992). Findings indicate self-efficacy influences an individual's choice among activities, persistence when problems arise (Bandura, 1986), and a variety of work outcomes, including performance ratings (Gardner & Pierce, 1998) and attendance (Latham & Frayne, 1989). Bandura (1989) asserted that people who believe they are capable of controlling their lives are more likely to be able to do so. Self-efficacy expectations are not concerned with the skills individuals possess but what they think they can do with whatever skills they possess (Bandura, 1986). Bandura (1989) asserted that people who believe they are capable of controlling their lives are more likely to be able to do so. Self-efficacy expectations are not concerned with the skills individuals possess but what they think they can do with whatever skills they possess (Bandura, 1986). Concordant with the definition of self-efficacy, collective efficacy refers to members' perceptions of their group's competency (Bandura, 1986) or aggregated ability to successfully complete a designated task (Guzzo et at, 1993).

Individuals with a strong sense of self-efficacy are able to adapt emotionally to the life's challenges and difficult situations without becoming overanxious (Maddux & Lewis, 1995) and depressed (Davis-Berman, 1988, 1990). Older individuals are faced with a variety of major life changes and losses. Perception of self-efficacy or confidence is an essential aspect of their coping efforts to deal with the negative life events that are associated with psychological distress and depression (Holahan & Holahan, 1987a, 1987b). In addition, highly efficacious older individuals are more independent (Abler & Fretz, 1988; Wahl, 1991) and have more enhanced feelings of control over their environment (Rodin, Bohm, & Wack, 1982) than do older individuals who are less efficacious. According to Myers (1990), efficacy expectations play a key role in successful aging. Aging is not an illness but a life condition that bears investigation with different parameters such as physical activity and self-efficacy. National cross-sectional surveys of physical activity consistently report that physical activity declines with age and those older adults participate in less physical exercise than do younger adults (Stephens & Caspersen, 1994). Physical activity refers to "any bodily movement produced by the contraction of skeletal muscle that increases energy expenditure above the basal level" (U.S. Department of Health and Human Services, 1996a, p. 20). The self-efficacy in this respect may prove as beneficial to increase and maintain the confidence of the elderly people. Beside this psychological interventions in general as well as favorable policies of state/center government may prove asset for increasing their well-being and reducing psychological frailness.

Policies and Elderly Frailty Well-Being

Nationally, the older population is expected to double in numbers through the year 2030. Health care providers are challenged to develop new models of care delivery for this unique population. Because of their complex clinical presentations and needs frail elderly people require another approach than people who age without many complications. Several inpatient geriatric health services have proven effectiveness in frail persons. However, the wish to live independently and policies that promote independent living as an answer to population aging call for community intervention models for frail elderly people. The models such as preventive home visits, comprehensive geriatric assessment, and intermediate care qualify, but their efficacy is controversial, especially in frail elderly persons living in the community.

With the Dutch EASY care Study Geriatric Intervention Programme (DGIP), a model, was development that study effectiveness of problem based community intervention models in frail elderly people. DGIP is a community intervention model for frail elderly persons where the GP refers elderly patients with a problem in cognition, mood, behaviour, mobility, and nutrition. A geriatric specialist nurse applies a guideline-based intervention with a limited number of follow up visits. The intervention starts with the application of the EASY care instrument for geriatric screening. The EASY care instrument assesses (instrumental) activities of daily life, cognition, mood, and includes a goal setting item. Effects on functional performance health related quality of life, and career burden are studied in an observer blinded randomized controlled trial. 151 participants were randomized over two treatment arms--DGIP and regular care--using pseudo cluster randomization. These visits were planned three and six months after inclusion. Process measures and cost measures were recorded. Intention to treat analyses was focused on post intervention differences between treatment groups. The design of a trial evaluating the effects of a community intervention model for frail elderly people was presented.

The problem-based participant selection procedure satisfied; few patients that the GP referred did not meet our eligibility criteria. The use of standard terminology makes detailed insight into the contents of our intervention possible using terminology others can understand well (Melis, van Eijken, Borm, Wensing, Adang, van de Lisdonk, van Achterberg, Olde Rikkert, 2005). One another project that demonstrates successful outcomes for communities. A model implemented community-based geriatric case management for frail elderly citizens residing in a private home or in an assisted living facility. Conventional hands-on delivery was combined with the distance-based convenience of tele- health. The outcomes prove this model to be cost-effective while improving quality of life for enrollees (Duke, 2005). There are various psycho-social factors associated with well-being of elderly man and women. The gender also makes them psychologically vulnerable.

Gender Construction and Elderly Frailty

The studies related to gender and well-being has its own importance. Ghizala Khan and Akbar Hussain, (2001) in their paper "Sex Differences in Psycho-physical Effects of Psychotropic Drugs among Elderly" have examined sex differences on various characteristics representing the psycho-physical effects of drugs. The main findings of the study show that female senescent scored significantly higher than of the male senescent on 'calm', 'depressed', 'drowsiness', 'healthy', 'relaxed', 'secure', and 'weakness' characteristics representing the psycho-physical effects of drugs, whereas male senescent scored significantly higher than the female senescent on 'sadness' and 'sickly' characteristics. However, significant differences were not found between the mean scores of male and female senescents on the remaining eleven characteristics associated with psycho-physical effects of drugs.

A study of Kabitsis and Harahousou, (2001) on "The assessment of functional Fitness and Attitudes towards Physical Fitness of Greek Elderly People" state that the purpose of their study was to assess the functional fitness and the attitudes towards physical fitness of the elderly people. A battery of tests was administered to 100 pensioners (55 men and 45 women) between 60 and 82 years of age. The parameters measured were: flexibility, body agility, coordination, strength and endurance. The data for the attitude assessment were collected by personal interviews using a standardized questionnaire. Preliminary analyses, i.e. item analysis, reliability estimates and factor analysis were used to validate the attitudinal inventory. The main results were the following: (a) women performed significantly better than men in the flexibility and coordination tests; (b) there was a significant difference between active and inactive subjects in the strength and endurance tests only, (c) the subjects of the younger group performed significantly better than subjects of the older group in all but the flexibility and endurance tests; (d) men had significantly more positive attitudes towards physical fitness than women; and (e) the subjects of the younger group had significantly more positive attitudes towards physical fitness than the subjects of the older group. ANOVA showed significant effects of the age and sex parameters only.

In a classic study of Chadha and Easwaramoorthy (2001) on "The Leisure activities and Indian Elderly" point out that the graying of the Indian population along with its fast pace of industrialization and urbanization is bringing about new changes and challenges for the Indian researchers. Research areas/concepts once considered as a western requirement are emerging as the need of the Indian society. One such is leisure time activities of Indian elderly. In this paper, an attempt has been made to provide a synthesis of some of Indian research work's carried in the area of leisure time activities which is strongly and positively related to the general wellbeing in the Indian elderly. Differences in leisure time activities were observed with reference to sex, marital status, rural-urban residence, occupation and pre-and post-retirement periods.

Through the devaluation of people's former life, i.e., their accumulated cultural capital and habitus-the modern society denies the value of the cultural capital of elderly people. By this devaluation the symbolic power of elderly people as a group diminishes. The present mechanisms of symbolic capital tend to legitimize mainly the work-related, active, midlife positions and activities. They also seem to emphasize the value of social capital leaning only on these qualities. Thus, they operate as lessening the social capital of old people. Their own social activities and networks are loosing their symbolic value. Through these processes the elderly population becomes easily divided by gender: the activities and networks typical to women's life course are undermined, the retirement may become too big a change to men.

LIFE SATISFACTION AMONG THE SENIOR CITIZEN

Attach your thoughts,
Old age, to Paradise:
Its perfume will rejuvenate you,
Its breath will give you youth,
Your rags will be hidden
By its splendid clothes
For you it drew this picture in Moses:
His wrinkled cheeks shone like the sun,
Signs that old age will find rejuvenation in Eden.

Most of elderly consider frailty as a curse whereas there are some other who bounce back its also static load, maintain homeostatic and resilience while shriving processes. Frailty has long been deemed to be an inevitable consequence of ageing. Despite this many other some have thought of frailty as a state of pre-death, followed with disability and co-morbidity (Pope & Tarlov, 1991; Hoffman, Rice, & Sung, 1996). There is clearly no consensus on the definition of frailty and thus is an emerging, controversial and enigmatic concept. The emergence of frailty as an increasingly important concept prompts inquiry into what it is. Is it a disease? Is it just ageing? Yet some younger adults are deemed frail and some older ones are not. Gillick (2001) observed that "frailty is a syndrome in desperate need of description and analysis. Yet, many papers have highlighted the importance of frailty in its relation to adverse outcomes which include dependency, disability, institutionalization, falls, injuries, acute illness, hospitalization, slow or blocked recovery and mortality (Hogan, MacKnight, & Bergman 2003). Frailness is the decrease in body's resiliency and its ability to compensate for stress during age related physiological changes. Frail refers to identity crisis in old age. Life expectancies for men and women are increasing (U.S. Bureau of the Census, 1997), and the over-65 age group is projected to double in size by 2020 (Bokovoy & Blair, 1994) and over will be quadrupled (Zedlewski, Barnes, Burt, McBride, & Meyer, 1990). Unprecedented degree of population growth continues has its serious repercussion on it as projected (U.S. Bureau of the Census, 1992).

Frailty and disability were often used interchangeably. There were a growing number of researchers who postulated that frailty was another term for disability in the elderly. By the 1990s, researchers felt that equating frailty with disability was inadequate. A number of difficult questions were asked. Were all older disabled patients frail? If not, why weren't they? How did one end up frail? Was it inevitable? What were the underlying mechanisms? Was frailty preventable? Is it clinically advantageous to delineate frailty? Current research is attempting to delineate the components of frailty (how do we identify it?) and the underlying physiological and biological mechanisms important in the development of a frail state but psychological mechanism remains dormant. Frailty is a dynamic and complex process with multiple interacting components.

The life course approach provides an attractive framework towards understanding frailty and its determinants. It integrates biological, social, clinical, cognitive, psychological and environmental factors, which interact across a person's life span. Studies on risk factors for frailty support this approach. There appeared various risk factors in mid to late life such as cognitive impairment, depression, disease burden, increased/decreased BMI, lower extremity function limitation, decreased social contacts, low physical activity, no compared to moderate alcohol consumption, poor perceived health, smoking and vision impairment as well as for functional decline. There emerged an evidence for an association among biological, psychological and social risk factors. This framework was modeled to set the stage for possible prevention/delay of the onset of frailty and its adverse outcomes. Observational studies on aging suggest associations between several lifestyle factors (exercise, nutrition, education, socioeconomic status, social/intellectual activities) and the onset of frailty. These findings provide opportunities for the development of interventions to promote healthy ageing, reduce the incidence of frailty, delay its onset and/or reduce the number of years in dependency (Vita, Terry, Hubert, & Fries, 1998)

The quality of life, dignity and personal well-being must always be the focus of the care and support someone receives. Social care needs to focus on outcomes - people may receive care which meets their physical needs, but their emotional and mental wellbeing are too often overlooked completely. In some care homes there is little understanding of dementia or depression, meaning that both conditions go unrecognized and untreated. The emotional exhaustion, is believed to be at the core of burnout (Maslach, 1982), and is characterized by a lack of energy and a feeling that one's emotional resources are used up (Cordes & Doughtery, 1993). Researchers have linked burnout to a variety of mental and physical health problems (Burke and Deszca, 1986; Maslach

& Pines, 1977), including decreased self-esteem, increased levels of irritability, depression, and anxiety, as well as the deterioration of family and social relationships (Jackson & Schuler, 1983).

This challenge was first highlighted in 1990. An American Medical Association white paper concluded that "one of the most important tasks that the medical community faces today is to prepare for the problems in caring for the elderly in the 1990s and the early 21st century (AMA, 1990). The report particularly emphasized the growing population of "frail, vulnerable older adults". It is estimated that in the community, *10 to 25% of those above 65 years are frail and 46% of those above 85 years are also frail.* In Singapore, the medical community is also faced with similar problems in caring for an ageing population. The proportion of older persons above the age of 65 was 6.8% of the population in 1995 and is projected to increase to 20% by the year 2030(The National Survey of Senior Citizens in Singapore, 1995). The increase in life expectancy and better healthcare would introduce an emergence of the very old and frail populations. Understanding the underlying mechanisms of the frail state points to opportunities for health promotion, prevention and interventions aimed at either delaying the onset of frailty or reducing its adverse outcomes.

The history of its official use dates back to the 1970s. Monsignor Charles F Fahey and the Federal Council on Aging in the United States were credited with introducing the "frail elderly" to describe a particular segment of the population (Maddox, 1987; Albert, 1995). This happened when the heterogeneity of the older population became more accepted. The term was selected to focus attention on a group of elderly with physical debilities, emotional impairments and debilitating social and physical environments. It was used as an administrative device for triggering access to core support services for the elderly. In the 1980s, researchers began to explain the term. Early definitions would include those aged 75 and older; a vulnerable population of seniors due to physical and mental impairment; individuals admitted to a geriatric programme; institutionalization; and those dependent on others for activities of daily living. (Streib, 1983; Reid, Gallagher, & Bosworth, 1986; Williams, Wynne, Woodhouse, & Rawlins, 1989). Chronic disease and its sequelae were felt to be the cause of increased vulnerability and frailty (Woodhouse, Wynne, Baillie, James, & Rawlins, 1998). For the human brain, there's no such thing as over the hill. Psychologists researching the normal changes of aging have found that although some aspects of memory and processing change as people get older, simple behavior changes can help people stay sharp for as long as possible. Frailty thus is an emerging, controversial and enigmatic concept needs further studies. Although there may be debate on the specific meaning of this term, there is no doubt about the impact of frailty on older individuals, their families and society by its adverse outcomes.

Health Beliefs

The health is not simply the absence of disease but something more positive, a joyful attitude towards life, and a cheerful acceptance of the responsibilities that life puts upon the individual. It is a state of complete physical, mental and social well being and not merely an absence of disease or infirmity. The concept of perfect and positive and health is a utopian creation of the human mind. It cannot become reality because man will never is so perfectly adapted to his environment that his life will not involve struggle, failure and sufferings. The healthy individuals are on who have a harmonious relationship with the community surroundings and the supernatural world. The health behaviour in rural and urban locality may also differ in men and women. It needs though analysis in below mentioned paragraph attempt has been made for the same in shimla district H.P.

Table 1 Average Score of Rural and Urban Elderly People On Physical Aspects Of The Neuroticism

Gender	Rural	Urban	Average
MALE	8.8	11.57	10.18
FEMALE	7.97	9.77	8.87
AVERAGE	8.38	10.67	9.52

From Table 1, it is quite clear that the average score of male on physical aspects of neuroticism was found (10.18) as compare to their counterpart female (8.87) regardless of locality. From the score it is quite evident that the male reported more physical health problem as compared to the female. Similarly average score of urban male was (11.57) whereas the average score of rural elderly male was (8.8). It shows that the urban male reported more physical health problem as compared to the rural male. Further the average score of urban female was (9.77) and rural female as (7.97). From the average score it is quite clear that urban female reported more physical health as compared to the rural female. Similarly average score of the rural elderly people was (8.38) whereas the average score of the urban elderly people was (10.67). It depicts that the urban elderly reported more physical health problem as compared to the rural counterpart. The scores have been shown in the figure .1a.

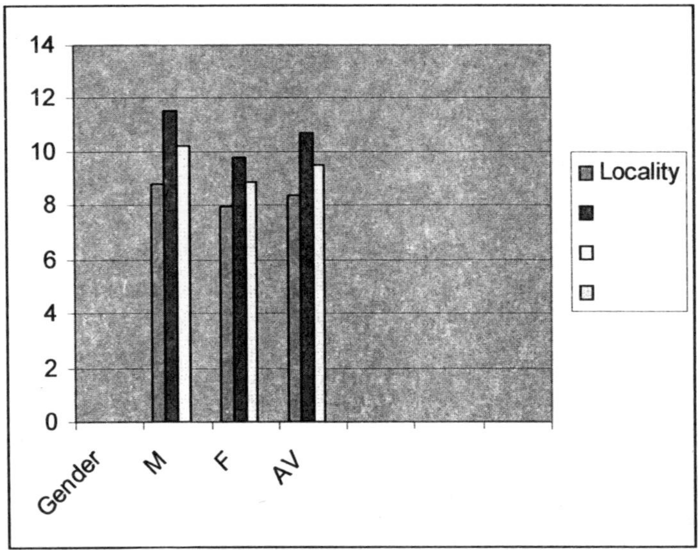

Fig. 1a Profile of rural and urban elderly people on Physical Aspects of the neuroticism.

Average score of rural male was found (8.8) as compared to the rural female (7.97). From the score it is quite clear that rural male reported more physically health problem as compared to rural female. Similarly average score of the urban male was (11.57) and urban female as (9.77). It shows that the urban male have been suffering more physical problems as compared to the urban female. The average score of rural male was (8.8) whereas the average score of urban female was (9.77). It shows the urban female having more physically health problem as compared to the rural male. Finally average score of rural female was (7.97) whereas the average score of urban male was (11.57). It also shows that the urban elderly male suffers more from physical health problem as compared to the rural female elderly people.

The aging perhaps is an integral part of human condition involves sub cultural phenomenon therefore is innovation of early nineteenth century (Atal, 2001). According to society theory Banton, Clifford, Erosh, Lousada and Rosenthal (1985) the societies enters into the life of an individual and construct him socially through economic, political and ideological practices in result can make him well or frail. Like Adam everybody in the unversed glow, progress and perform better in their youth and lost everything in the old age. “Old age”, the very word brings to mind white hair, stooped shoulders, an uncertain gait, & certain isolation. No one knows when old age begins. The process of ageing varies with individuals; some begins to look old at 59, while other looks young even at 65 or 70 years. Ageing is the universal process. Sometimes factors like low income, reduced physical mobility due to illness, or lack of appropriate transport can leave some people at home alone with nothing much to do. It also speaks to the isolation of researchers working on frailty (Hogan, MacKnight, & Bergman, 2003). Various genetic and environmental factors and no two individual ages in the same way (Miller, 2003). There is unobservable heterogeneity of factors behinds the frailty (Suchindran & Koo, 1997).

Table 2 Average Score On Rural And Urban Elderly Population On Psychological Aspects Of The Neuroticism.

Gender	Rural	Urban	Average
MALE	8.43	9.03	8.73
FEMALE	9.84	12.03	10.93
AVERAGE	9.13	10.53	9.83

From Table 2, it is quite clear that the average score of male on psychological aspects of health was found (8.73) as compare to the counterpart female (10.93) regardless of locality. From the score it is quite evident that the female are suffering from reported more mental health problem as compared to the male. Average score of urban male was (9.03) whereas the average score of rural male was (8.43). It shows that the urban male suffers from more mental health problem as compared to the rural male. Similarly the average score of urban female was (12.03) and rural female was (9.84). From the average score it is quite clear that urban female reported more psychological problem as compared to the rural female. Average score of the rural elderly people was (9.13) as compare to the urban counterpart (10.53) depicts that the urban elderly reported more mental health problem as compared to the rural elderly people.

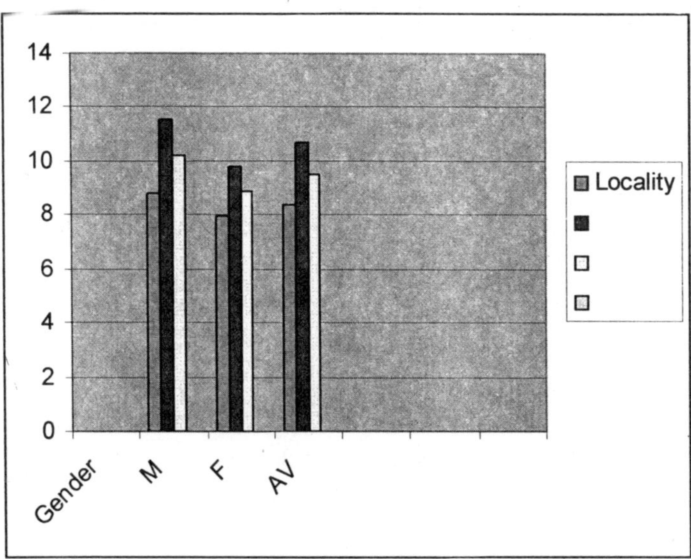

Fig.2b, Profile of rural and urban elderly people on psychological aspects of the neuroticism.

Average score of rural male was found (8.43) whereas average score of rural female was (9.84). From the score it is quite clear that rural female more mental health problem as compare to male. Similarly average score of the urban male was (9.03) and urban female was (12.03). It again shows that the urban elderly female have been suffering from psychological problems as compared to the counterpart living in Shimla and surrounding areas. The average score of rural male was (8.43) whereas the score of urban female was (12.03). It shows the urban female having more mental health problem as compared to the rural elderly male. The average score of rural female was (9.84) whereas the average score of urban male was (9.03). It shows that the rural females have slightly more psychological problem as compared to the urban male.

Many measure of health needs to assess social and emotional aspects of health as well as assessing presence or absence of disease (Wilson & Cleary, 1995). Same is the life of the elderly population which is influenced by various types of biological, social, psychological and cultural factors those make them psychologically frail by affecting their wellbeing. According to a conservative estimate roughly there are 100,000 people worldwide those die each day of age-related causes (Aubrey & de Grey, 2007). The emotional exhaustion, is believed to be at the core of burnout (Maslach, 1982), and is characterized by a lack of energy and a feeling that one's emotional resources are used up (Cordes & Doughtery, 1993). Researchers have linked burnout to a variety of mental and physical health problems (Burke and Deszca, 1986; Maslach and Pines, 1977), including decreased self-esteem, increased levels of irritability, depression, and anxiety, as well as the deterioration of family and social relationships (Jackson and Schuler, 1983) that in turn can be associated with psychological frailness.

Table 3 Average Score of Rural and Urban Elderly Population on Physical and Psychological Aspects of Neuroticism.

Gender	Rural	Urban	Average
MALE	18.63	17.53	18.08
FEMALE	17.80	21.80	19.80
AVERAGE	18.21	19.66	18.93

From Table 3, it is quite clear that the average score of male was found (18.08) as compare to the counterpart female (19.80) regardless of locality. From the score it is quite evident that the female reported more health problem as compared to the male. Average score of urban male was (17.53) whereas the average score of rural male was (18.63). It shows that the rural male reported more health problems as compared to the urban male. Similarly the average score of urban female was (21.80) and rural female was (17.80). From the average score it is quite clear that urban female reported more health problem as compare to the rural female. Average score of the rural elderly people was (18.21) whereas average score of urban elderly people was (19.66). It depicts that the urban elderly reported more health problem as compared to the rural counterpart. Average score of rural male was found (18.63) whereas compare to the rural female was (17.80). From the score it is quite clear that rural male was more health problem as compare to urban female.

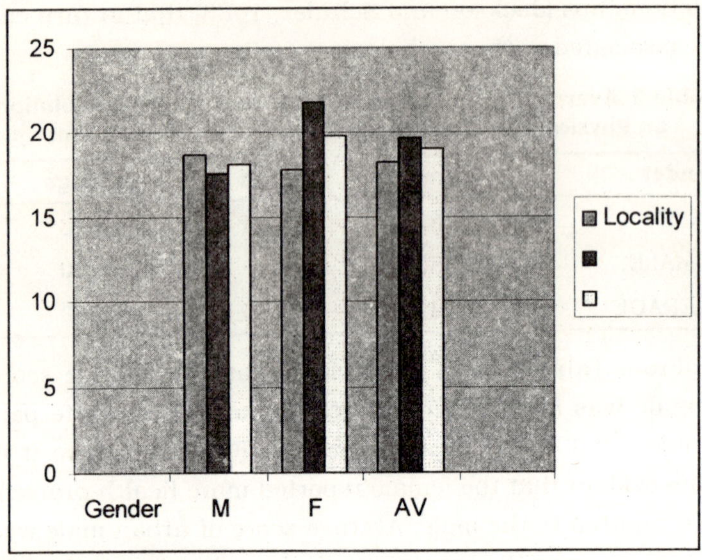

Fig.3c, Profile of rural and urban people on physical and psychological aspects of the neuroticism.

Similarly average score of the urban male was (17.53) and urban female was (21.80). It shows that the urban female suffering from health problems as compared to the urban female. The average score of rural male was (18.63) whereas the average score of urban female was (21.80). It shows the urban female having more health problem as compare to rural male. The average score of rural female was (17.80) whereas the average score of urban male was (17.53). It shows that the rural females have slightly more health problem as compared to the urban male.

Many factors such as demographic (accelerated ageing of the population), social (break-up of families, children moving away to find work), economic (low income women living alone), health (increased life expectancy, high incidence of disabilities) and financial (reduced health. Biological and non-biological factors are prevalent. According to Joseph, (2001) in his study "Influence of the old on the perspectives of the young on Human life" identifies interaction of the young with the old as influencing perspectives of the former on human life. Challam (1982) has stated that generally changes take place in four main spheres-physical, behavioural biological and intellectual.

Anxiety among the Senior Citizen

Anxiety is a psychological and physiological state characterized by cognitive, somatic, emotional, and behavioral components. These components combine to create an unpleasant feeling that is typically associated with uneasiness, apprehension, fear, or worry. Anxiety is a generalized mood condition that can often occur without an identifiable triggering stimulus. As such, it is distinguished from fear, which occurs in the presence of an observed threat. Additionally, fear is related to the specific behaviors of escape and avoidance, whereas anxiety is the result of threats that are perceived to be uncontrollable or unavoidable. Another view is that anxiety is "a future-oriented mood state in which one is ready or prepared to attempt to cope with upcoming negative events suggesting that it is a distinction between future vs. present dangers that divides anxiety and fear. Anxiety is considered to be a normal reaction to stress. More approximately the anxiety is the state of tension and apprehension that is a natural response to perceived threat. It may help a person to deal with a difficult situation, for example at work or at school, by prompting one to cope with it. When anxiety becomes excessive, it may fall under the classification of an anxiety disorder.

The anxiety is an emotional state, represented by a feeling of dread, apprehension, or fear. In humans, this can be defined by description using language; in animals, it must be inferred from behavioural observations. Tests of anxiety in man are thus based on self report, and these may be divided into features that characterize the person's temperament (*'trait' anxiety*) or that describe a current emotional state (*'state' anxiety*). In animals, it is inferred by the animal's response to an anxiety-provoking situation such as a threatening environment. Distinctions between anxiety and other emotional states, such as fear or even 'arousal', are not always clear. Also, there are close associations between cognition and emotion: man has the capacity not only to know, but also to respond emotionally to what he knows.

Table 4 Average Score of Rural and Urban Elderly Population on State Anxiety.

Gender	Rural	Urban	Average
MALE	38.00	38.17	38.05
FEMALE	44.63	39.17	41.90
AVERAGE	41.31	38.67	39.98

From Table 4, it is quite clear that the average score of male was found (38.05) as compare to their counterpart female (41.90) regardless of locality. From the score it is quite evident that the female reported more anxiety as compare to male. Average score of urban male was (38.17) whereas the average score of rural elderly male was (38). It shows that the urban male reported slightly more state anxiety as compare to rural elderly male. Similarly the average score of urban female was (39.17) and rural female was (44.63). From the average score it is quite clear that rural female reported more anxiety as compared to the urban female. Average score of the rural elderly people was (41.31) whereas average score of urban elderly was (38.67) depicts that the rural elderly people more anxiety than the urban elderly people.

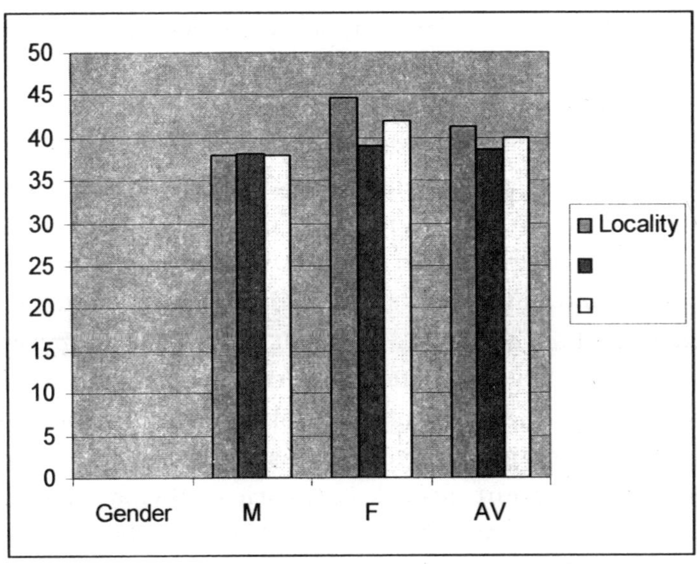

Fig.4 Profile of rural and urban elderly people on STAI anxiety.

Average score of rural male was found (38.00) whereas compare to the rural female was (44.63). From the score it is quite clear that rural female having more anxiety as compare to rural male. Similarly average score of the urban male was (38.17) and urban female was found (39.17). It also shows that the urban female having more anxiety than the urban male. The average score of rural male was (38.00) whereas the average score of urban female was (39.17). It is quite clear that the urban female again reported more anxiety as compare to the rural male. The average score of rural female was (44.63) whereas the average score of urban male was (38.17). It also shows that the rural elderly female having more anxiety than the urban elderly male.

According to Malher, Director General World Health Organization (1982), "Ageing" is not simply a physical process but a state of mind and today we witnessing the beginning of a revolutionary change in that state of mind. For biologist and medical scientists ageing refers to deterioration in psychological capabilities. According to Cowdry (1942) the changes found in aged persons as structural alteration due to infections, toxins, traumas and nutritional disturbances or inadequacies give rise to what are called degenerative changes and impairments. Challam (1982) has stated that generally changes take place in four main spheres-physical, behavioural biological and intellectual. Empirical studies have indicated that definitions relating the ascription of old age as well as the characteristics associated with the aaged vary in different social categories. Swiss scholars Vollenwyder, Bickel, d'Epinay and Maystre (2001) in their paper "changing Family Networks and Relationships in Two Swiss cohorts of elderly (1979-1994)" have analyzed the changes occurring in the structure of the family network of older people and their family interactions over a 15-year period

Life Satisfaction among the Senior Citizen

The life satisfaction deals with the sense of well-being and may be assessed in terms of mood, satisfaction with relations with others and with achieved goals, self-concepts, and self-perceived ability to cope with daily life. It is an overall assessment of feelings and attitudes about one's life at a particular point in time ranging from negative to positive.

It is one of three major indicators of well-being: life satisfaction, positive affect, and negative affect. Although satisfaction with current life circumstances is often assessed in research studies, it also includes the desire to change one's life; satisfaction with past; satisfaction with future; and significant other's views of one's life."

Table 5 Average Score of Rural and Urban Elderly Population on Life Satisfaction

Gender	Rural	Urban	Average
MALE	25.77	26.77	26.27
FEMALE	26.67	27.00	26.84
AVERAGE	26.22	26.88	26.55

From Table 5, it is quite clear that the average score of male was found (26.27) as compare to their counterpart female (26.84) regardless of locality. From the score it is quite evident that the female reported slightly more life satisfaction as compare to the male. Average score of urban male was (26.77) whereas the average score of rural elderly male was (25.77). It shows that the urban male reported more life satisfaction than the rural male. Similarly the average score of urban female was (27.00) and rural female was (26.67). It also shows that the urban females have been more satisfied with their life as compared to the rural female. Average score of the rural elderly people was (26.22) whereas average score of urban elderly people was (26.88) depicts that the rural elderly reported more life satisfaction as compare to rural elderly people.

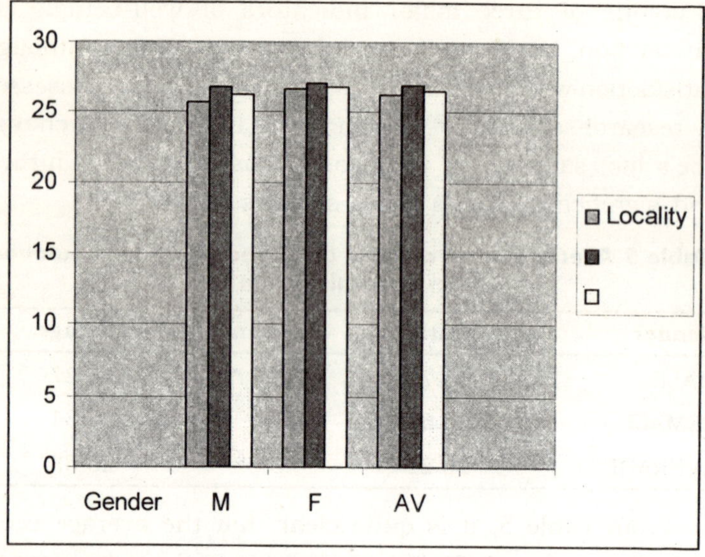

Fig. 5 Profile of rural and urban elderly people on Life Satisfaction.

Average score of rural male was found (25.77) as compare to the rural female (26.67). From the score it is quite clear those rural females are more satisfied with their life as compare to the rural male. Similarly average score of the urban male was (26.77) and urban female was (27.00). It also shows that the urban females are more satisfied with their life as compared to urban male elderly people. Finally the average score of rural male was (25.77) whereas the average score of urban female was found (27.00). It shows that the urban female elderly people were more satisfied with their whole life as compare to the rural elderly male. Similarly the average score of rural female was (26.67) whereas the average score of urban male was found (26.77). It also shows that the urban males are more satisfied with their whole life as compared to the rural females. According to mostly the old people are less satisfied with increasing age. The depression is reported in around a third of residents and dementia in two thirds (Dening, & Milne, 2009). Many researchers have have used frailty synonymously with disability and co-morbidity (Pope & Tarlov, 1991; Hoffman, Rice, & Sung, 1996). Large scale individual based longitudinal data should be studied in order to gain more insight into the relative importance of the geographical drift hypothesis (**Verheij, 1996**). Like Adam everybody in the unversed glow, progress and perform better in their youth and lost everything in the old age. "Old age", the very word brings to mind white hair, stooped shoulders, an uncertain gait, & certain isolation. No one knows when old age begins. The process of ageing varies with individuals; some begins to look old at 59, while other looks young even at 65 or 70 years. Ageing is the universal process. Human life from conception to death is a complete sequence of events. Human concern about the phenomenon of ageing is age old. Disengagement theory as proposed by Cumming and Henery (1961) which states that individual prefers to give up his earlier activities and social roles and society tends to take away his earlier professional and social responsibilities both not minding but mutually and willingly seeking the disengagement. Maddox (1964) emphasizes on psychological and social withdrawal model and states that disengagement process is intrinsic and inevitable is disputed and there are factor such as sex, health, profession, intelligence, personality type and life style which modify the disengagement process. According to Seal (1979) elderly people suffer with national, special and personal problem. In the word of Rao (1979) the aged people suffer with socio-economic condition like poverty, breaking up of joint family system and poor service for the aged pose a psychiatric threat to them.

Further a study of Ramamurti (1970) observed that life satisfaction declined significantly around retirement period. A large number of studies have been conducted on adjustment (Ramamurti, 1968a, Siddaiah, 1970; Desai & Naik, 1970;Paintal, 1970; Shalini Bhogle et al, 1978, Anantharaman, 1976, Chandrika & Anantharaman, 1982; Hrudayanatha & Reddy, 1984; & Menachery, 1985). Beside this other variable inld age are also causing rigidity and inflexibility (Ramamurthi, 1970), husband wife communication (Jamuna and Ramamurti, 1984), marital satisfaction (Prakash,s 1983), attitude toward future and death (Anantharaman, 1980), role activity (Jamuna, 1984), family type and socioeconomic status (Desai & Naik, 1970) are related to better adjustment.

CONCLUSION

It is the tendency of every society to live and to go on living. But the society extols the strength and the fecundity that is so closely linked with youth and dreads the worn-out sterility, the decrepitude of age. Since the dawn of history old people have regretted their youth and young people have feared the onset of old age. According to western thought, old age is an evil, an infirmity and a dreary time of preparation for death they are often envisaged with more sympathy than is decrepitude, because death means deliverance. The situation of old people expresses the ambiguity of the human condition more fully than do the other ages of life. They are condemned to veneration or detestation, and no longer have the right to commit the slightest mistake. He who enjoys so much experience, he can no longer surrender to the slightest urge of the flesh, he who is so worn and shriveled, he must be perfect, or he will become revolting and doting. The objective of the present study was to assessing the psychological frailness among rural and urban elderly residents of Shimla District. More specifically the objective of the present study was to study the psychological frailness among rural and urban elderly resident of Shimla. For accomplishing the objective the study was conducted on a sample of N = 120 elderly population those were divided into two comparable halves based on their locality that comprises of N = 60 a rural and remaining N = 60 in urban areas. Finally the subjects were divided into two comparable halves based on gender that comprises of N = 30 subjects in each group. In all there were four groups with N = 30 subjects in each selected from rural and urban areas of Shimla. In all there were four groups with N = 30 subject in each that comprises of a total sample of N = 120 subjects. More specifically the subjects were selected from Mashobra, Chamyana, Malyana and Sangti villages as well as from Shimla town, though purposive sampling of Sanjauli A 2 x 2 factorial designed was used. Further both qualitative and quantitative analysis was made for assessing the subjects more deeply and reliably to report psychological frailness. In qualitative analysis field observation, Interview, Fuzzy Cognitive Mapping and photography were used. In quantitative measures PGI Health Questionnaire, State Anxiety Inventory and satisfaction with life scale was used. Since the researcher herself live in Shimla town, therefore helped a lot for carrying out in making observation on the locality of this town for this manuscript. The researchers asked various people in the town order to understand and record the psychological frailness in elderly people. Various Gram Panchayat Pradhan, educated youths as well as elderly people were asked to know deeply about the psychological problem and emotional weakness in the old people, living in Shimla and surrounding villages. The researcher entered in the particular village, by establishing warmth atmospheres and addressing the objective of the study to the elderly. The ideas elicited by the subjects was welcomed by the researchers and they were further encouraged to explain more about their physical, mental and social problems those were formed to be involved in their psychological vulnerabilities.

The data was tabulated and analyzed with the help of two way analysis of variance. The result of PGI Health questionnaire Area-A revealed that the only main effect of locality on physical aspects of the neuroticism was found F (1,119) = 2.98, p <.05 as statistically significant. The two way interaction between locality and gender was also found F (1,119) = 2.98. p <.05 statistically significant. Urban elderly showed more physical health problem as compare to their rural counterpart. Further elderly male reported more physical health problem as compare to the female.

In psychological part of the neuroticism, the main effect of locality F (1,119) = 6.73, p <.01 and gender was found F (1,119) = 16.61, p>.001 as statistically significant. None of the interaction emerged as significant. More appropriately the urban elderly reported more psychological problem as compared to rural one. Female reported more psychological problem than to elderly male. In both physical and psychological aspects of the neuroticism the main effect of locality was found F (1,119) = 1.63, p >.05 as statistically non-significant. Beside this the two way interaction between locality and gender was also found F (1,119) = 5.04. p <.01 as statistically significant. Urban elderly reported more health problem as compare to rural elderly counterpart. Female reported more psychological problem as compare to male. **Similarly in state anxiety** the only main effect gender was found F (1,119) = 5.53, p<.01 as statistically significant. None of the interactions emerged as significant. While using qualitative analysis the elderly people living in rural as well as in urban area reported deep pain. The old people reported physical, social and psychological frailness that includes such as lack of energy, dizziness, and feeling of pressure, family member torture, aggression, their low social support, poor sense organ, hallucination and delusion, restlessness, guilty, suicide feeling, rheumatism, trembling / shivering, feeling of unworthy, irritable, lost interest in things, death anxiety and related fear, livelihood problem, pocket money, problem and homeostasis problem. Beside this they reported extreme pressure on their mind due to seclusion widowed, illness and poor attitude towards ageing, and life the treatment they receive from their family, younger population as well as higher and respective communities. They also found apprehensive about their changing physique, disease, decay and activities. The loneliness/isolation problem followed with existential sadness and social status followed with lack of transportation, recreational and other sitting avenues in city and village, walking and stepping up of in stares made them apprehensive. They reported little life satisfaction and most of the god to be cherished. Most of the elderly people reported adjustment problem. They also reported recent life stressful event associated with their poor well-being. The death of their loved one, socio- economic problem, unrecognizing of their wisdom and thought as well as argument of younger group and then non-co-operative attitude towards their frailty. They also reported digestion problem, shelter and bedding problem.

Most of the people reported that they are not given proper cots; beds (Razai) and gaddas for sleeping therefore feel extreme cold and wake up early in the morning. Besides this bathing, clouting problem make then vulnerable toward unhygienic condition and diseases causing their health problem. Most of their family and son as well as daughter in law consider then as burden on their economy. Their emotional needs are not properly satisfied. Besides this racial differential in communities make them vulnerable. The accidental tragedies such as widowhood and loneliness and without family support followed with poor social support make their vulnerable. Not only this poor understanding and poor welfare programme and policies towards elderly people in rural and urban areas further aggravate their problem.

SUGGESTIONS

1. Invite your loved one out. Depression is less likely when people's bodies and minds remain active. Suggest activities to do together that your loved one used to enjoy: walks, an art class, a trip to the museum or the movies—anything that provides mental or physical stimulation.
2. Schedule regular social activities. Group outings, visits from friends and family members, or trips to the local senior or community center can help combat isolation and loneliness. Be gently insistent if your plans are refused: depressed people often feel better when they're around others.
3. Schedule regular social activities. Group outings, visits from friends and family members, or trips to the local senior or community center can help combat isolation and loneliness. Be gently insistent if your plans are refused: depressed people often feel better when they're around others.
4. Connecting to others – Limit the time you're alone. If you can't get out to s Plan and prepare healthy meals. A poor diet can make depression worse, so make sure your loved one is eating right, with plenty of fruit, vegetables, whole grains, and some protein at every meal.

5. Plan and prepare healthy meals. A poor diet can make depression worse, so make sure your loved one is eating right, with plenty of fruit, vegetables, whole grains, and some protein at every meal.
6. Taking care of a pet – Get a pet to keep you company.
7. Learning a new skill – Pick something that you've always wanted to learn, or that sparks your imagination and creativity.
8. Enjoying jokes and stories – Laughter provides a mood boost, so swap humorous stories and jokes with your loved ones, watch a comedy, or read a funny book.
9. Exercising – Even if you're ill, frail, or disabled, there are many safe exercises you can do to build your strength and boost your mood—even from a chair or wheelchair.
10. Encourage the person to follow through with treatment. Depression usually recurs when treatment is stopped too soon, so help your loved one keep up with his or her treatment plan. If it isn't helping, look into other medications and therapies.
11. Schedule an appointment with a doctor for a thorough evaluation, including a complete physical and lab workup. This is particularly important since many medical conditions, medications, and even certain physiological changes of aging can cause depression or compound the problem.
12. Watch for suicide warning signs. Seek immediate professional help if you suspect that your loved one is thinking about suicide.
13. Provide assistance in Pension for old man; and pension after spouse death.

14. Provide facilities of dieting; game and park facilities; visual hearing aids; aids for physically handicapped, free travelling in buses, rail, airways, sea etc as well as provide bench facilities in local pedestal and bazaar.
15. Govt. advocate for fighting any civil and criminal suits be provided o them
16. To prevent harassment against elderly Panchayat Pradhan be given power to punish the notorious element.
17. The scheduled castes and tribe people including migrated and displaced elderly be given prior importance for rehabilitation as well as providing free grocery or rashan facilities, bus/train/air journey.
18. Free clothes and shoes free facilities followed with education be provided to them.
19. Some nutrition such meal facilities as kaju, badam, milk, or other dry fruits be available at community centre packed after grinding.
20. After retirement the institution should pay full salary to the old people.
21. Elderly people's communication with family member be encouraged.
22. Death anxiety is reduced and optimistic attitude husband wife communication.
23. Describe, explain, and produce behavior change interventions with a broad range of health-related behaviors.
24. The active you are physically, mentally and socially better you will feel.

REFERENCES

Adammier, I. M. & Baltes, P.P. (1972). Objectives Vs. Perceived differences in presently. *Journal of Gerontology, 21,* 46-51.

Adler, A. (1964). *Superiority and social interest: A collection of later writings.* New York: University Press.

Ahammier, I.M. & Batles, P.P. (1972). Objectives Vs. perceived differences in personality. *Journal of Gerontology, 21,* 46-51.

Albert, D.M.(1995). *Profiles in gerontology: A biographical dictionary.* Westport, Connecticut: Greenwood Press.

Aminoff, M.J.(2000). *Brown-Sequard – A visionary of science.* New York: Raven Press.

Anantharaman, R.N. (1976). Psychology of ageing: A study in adjustment. *Ph.D thesis,* Bangalore: University of Bangalore.

Ansbacher, H. L. (1991). The concept of social interest. *Individual Psychology, 47,* 29-46.

Atal, Y. (2001). The United Nations and ageing. In I.Modi (ed.), *Ageing human development (pp. 1-9).* New Delhi: Rawat Publication.

Atchley, R. (1976). *The sociology of retirement.* New York: Halsted Press.

Aubrey, D.N.J & de Grey (2007). "Life span extension research and public debate: societal considerations" (PDF). *Studies in Ethics, Law, and Technology,* 1, 1-5.

Bandura, A. (1977). Self-efficacy: Toward a unifying theory of behavioural change. *Psychological Review 84,* 191-215.

Bandura, A. (1977). Self-Efficacy: Toward a unifying theory of behavioural change. *Psychological Review 84,* 191-215.

Bandura, A. (1978). Reflections of self-efficacy. In S. Rachman (ed.), *Advances in behaviour research and therapy* (pp. 237-269). New York : Freeman and Company.

Bandura, A. (1986). *Social foundations of thought and action: A social cognitive theory.* Englewood Cliffs, NJ: Prentice-Hall.

Bandura, A. (1997). *Self-efficacy: The exercise of control.* New York, NY: W. H. Freeman and Company.

Bhawsar, R.D. (2001). Population ageing in India: Demographic and health dimensions. In I.Modi (ed.), *Ageing human development (pp. 256-277).* New Delhi: Rawat Publication.

Bowen, R. & Atwood, C.S. (2004). Living and dying for sex. A theory of aging based on the modulation of cell cycle signaling by reproductive hormones. *Gerontology, 50 (5),* 265–90.

Bowling, A. (1995). *Ageing well: Quality of life in old age.* Milton Keynes: Open University Press.

Brocklehurst, J.C.(1985). *The day hospital. In textbook of geriatric medicine and gerontology, 3rd edition.* London: Churchill.

Bromley, D.B. (1974). *The Psychology of human aging.* London: Penguin Books Ltd.

Calabrese, E.J. & Baldwin, L.A.(1985). Defining hormesis. *Human Exp Toxicol, 21,* 91-7.

Chadha, N.K. & Easwaramoorthy, M. (2001). Leisure time activities and Indian elderly. In I.Modi (ed.), *Ageing human development (pp. 374-380).* New Delhi: Rawat Publication.

Challam, R. (1982*).* Science of gerontology. *Economic Times, 30th May,* 1-5.

Chen, P.C. (1994). *Psychosocial factors and the health of the elderly Malaysians.* Ann-Acad-Med-Singapore, 16(1), 110-4.

Chen, S. & Lu, Y.E. (2001). Community care and social support for older persons in UK, USA and China: An international perspective. In I. Modi (ed.), *Ageing human development (109-125).* New Delhi: Rawat Publication.

Cohen, H.J. (2000). In search of the underlying mechanisms of frailty. *J Gerontol Med Sci, 55A,* 706-8.

Cordes, C.L. & Doughtery, T. W. (1993). A review and an integration of research in job burnout. *Academy of Management Review, 18,* 621-656.

Cowdry, E. V. (1942). *Problems of Aging.* Bolbimore: Williams & Wilkins.

Crawford, K. (1996). Vygotskian approaches to human development in the information era. *Educational Studies in Mathematics,* (31), 43-62.

Darshinamuriti, S. (1962). *Introduction to preventive and social medicine (Vol-I) (1st Ed)* Kokinda: Published by education enterprises.

Davidson, K. (2001). Reconstructing life after a death: Psychological adaptation and social role transition in the medium and long term for older widowed men and women in the UK. In I.Modi (ed.), *Ageing human development (pp. 221-236).* New Delhi: Rawat Publication.

de Castro, L.R. & de Castro, G. R. (2001). Coping in old age: considerations from the point of view of care studies from Brazil. In I.Modi (ed.), *Ageing human development (pp. 181-197).* New Delhi: Rawat Publication.

de Oca, V.M. (2001). Discourses, voices and visions on ageing in Mexico City. In I.Modi (ed.), *Ageing human development (pp. 53-66).* New Delhi: Rawat Publication.

Dening, T & Milne, A., (2009). *Depression and mental health in care homes for older people.* Britinnika

Desai, K.G. & Naik, R.D. (1971). *Problems of the retired people in Greater Bombay:* In H.S Bhatia (ed), Ageing and society, Udaipur: Arya Book Center.

Dixon, A.L. S. (2001). Cross culturally comparative structural affecting the social aspects of ageing. In I. Modi (ed.), *Ageing human development (pp. 10-29).* New Delhi: Rawat Publication.

Dube, S. C. (1975). Introduction. In H. S. Parmar (Ed.), *Polyandry in Himalaya (pp. ix-xii).* Delhi: Vikas Publishing House Pvt. Ltd.

Duke, C. (2005). *The frail elderly community-based case management project.* Geriatr Nurs., 26(2), 122-127.

Dutta, C.(1997). Significance of sarcopenia in the elderly. *J Nutr, 127,* 992-993.

Elango,S.(1998). A study of health and health related social problems in the geriatric population in a rural area of Tamil Nadu. Indian *Journal of Public Health, 42,* (7-8).

Fried, L.P., Tangen, C., Waltson, J., Newman, A.B., Hirsch, C. & Gottdiener, J. (2001). Frailty in older adults: evidence for a phenotype. *Journal of Gerontological Medical Sciences, 56,* 146-56.

Gardner, D. G. & J. J. Pierce. (1998). Self-esteem and self-efficacy within the organizational context: An Empirical examination. *Group and Organization Management, 23,* 48-70.

Gastron, L., Andres, H. & Vuyosevich, J. (2001). Ageism at school: images and stereotypes of ageing and the old age in Argentina. In I.Modi (ed.), *Ageing human development (160-180).* New Delhi: Rawat Publication.

Ghosal, B.C. (1982). Care of the aged. *Yojana, 26,* 10.

Giard, N., Lichenstein, P. & Yashin, A.I.(2002). A multistage model for the genetic analysis of the ageing process. *Stat Med 21,* 2511-26.

Gillick M. (2001). Pinning down frailty. *J Gerontol Med Sci, 56A,* 134-5.

Gist, M. (1987). Self-efficacy: implications for organizational behaviour and human resource management. *Academy of Management Review, 12,* 472-485.

Gist, M. E., Mitchell, T. R. (1992). Self-efficacy: A Theoretical analysis of its Determinants and Melleability. *Academy of Management Review, 17 (2),* 183-211.

Grey, A. (1988). Sullivan's contributions to psychoanalysis. *Contemporary Psychoanalysis, 4,* 548-576.

Gurumurthy, K.G. (1998). *The age in India.* New Delhi: Reliance Publishing House.

Hallberg, I. R., & Kristensson J., (2004). Preventive home care of frail older people: a review of recent case management studies. *J Clin Nurs, 13(6B),* 112-20.

Havens, B. & Hall, M. (2001). Social isolation, loneliness, and the health of older adults in Manitoba, Canada. In I. Modi (ed.), *Ageing human development (pp. 126-144).* New Delhi: Rawat Publication.

Hébert, R., Durand, P. J., Dubuc, N., Tourigny, A., & The PRISMA Group (2003). *International Journal of Integrated Care.*

Hitt, R., Young-Xu, Y., Silver, M.& Perls, T.(1999). Centenarians: the older you get, the healthier you have been. *Lancet, 354,* 652.

Hoffman, C., Rice, D. & Sung, H.Y. (1996). *Persons with chronic conditions: their prevalence and costs. JAMA, 276,* 1473-9.

Hoffman, C., Rice, D. & Sung, H.Y. (1996). Persons with chronic conditions: their prevalence and costs. *JAMA 276,* 1473-9.

Hogan, D.B., MacKnight, C. & Bergman, H.(2003). Steering committee, Canadian initiative on frailty and aging. Models, definitions and criteria of frailty. *Aging Clin Exp Res, 15(3 Suppl.),* 1-29.

Hogan, D.B., MacKnight, C.& Bergman, H.(2003).Steering committee, Canadian initiative on frailty and aging. Models, definitions and criteria of frailty. *Aging Clin Exp Res, 15(3 Suppl.),* 1-29.

Hogan, D.B., MacKnight, C., & Bergman, H. (2003). Steering committeé, Canadian initiative on frailty and aging. Models, definitions and criteria of frailty. *Aging Clin Exp Res, 15(3 Suppl.),* 1-29.

Hossain, M.R. (2001). Demography of ageing and pattern of old age security in Bangladesh. . In I.Modi (ed.), *Ageing human development (pp. 73-81).* New Delhi: Rawat Publication.

Jamuna, D. (2001). Intergenerational issues in elder care. In I.Modi (ed.), *Ageing human development (pp. 287-295).* New Delhi: Rawat Publication.

Joseph, J.C. (2001). Influence of the old on the perspectives of the young on human life. In I.Modi (ed.), *Ageing human development.* New Delhi: Rawat Publication.

Kabitsis, C.N. & Harahousou, Y.S. (2001). The assessment of functional fitness and attitudes towards physical fitness of Greek elderly people. In I.Modi (ed.), *Ageing human development (pp. 237-243).* New Delhi: Rawat Publication.

Kesby, S. G. (2002). Nursing care and collaborative practice. *Journal of Clinical Nursing, 11(3),* 357-66.

Kessler, R.C. (2005). Prevalence, severity, and co morbidity of twelve-month DSM-IV disorders in the National Comorbidity Survey Replication (NCS-R). *Archives of General Psychiatry, 62*(6):617-27.

Kessler, R.C., Berglund, P.A., Demler, O., Jin, R. & Walters, E.E.(2005) Lifetime prevalence and age-of-onset distributions of DSM-IV disorders in the National Co morbidity Survey Replication (NCS-R). *Archives of General Psychiatry, 62(6),* 593-602.

Khan, G. & Husain, A. (2001). Sex differences in psycho-physical effects of psychotropic drugs among the elderly. In I.Modi (ed.), *Ageing human development (pp. 381-384).* New Delhi: Rawat Publication.

Kirkwood, T.B.(1989). Molecular gerontology. *J Inherit Metab Dis, 25,*189-96.

Kolata,G.(2004). Is frailty an inevitable part of aging? In: *International Herald Tribune (online).* www.iht.com/articles/77611.html

Kolata,G.(2004). Is frailty an inevitable part of aging? *International Herald Tribune (online): www.iht.com/articles/77611.html.*

Lamberts, S.W., van den Beld, A.W. & van der Lely, A.J.(1997). The endocrinology of aging. *Science, 278,* 419-24.

Lan.(2006). *The psychology of ageing: An introduction.* London: Jessica Kingsley Publishers.

Laslett, P. (1989). *A fresh map of life: the emergence of the third age.* London Weidenfold, & Nicholson.

Latham, G. P. & Frayne, C. A. (1989). Self management training for increasing job attendance: a follow-up and a replication. *Journal of Applied Psychology, 74,* 411-416.

Lebel, P., Leduc, N., Leclerc, C., Contandriopoulos, A.P., & Beland, F. (1999). Un modele dynamique de la fragilite. *L'Annee gerontologique, 13,* 84-94.

Linnane, A.W., Marzuki, S., Ozawa, T. & Tanaka, M.(1989). Mitochondrial DNA mutations as an important contributor to ageing and degenerative diseases. *Lancet,1,* 642-5.

Lipsitz, L.A & Goldberger, A.L. (1992). Loss of complexity and aging; potential applications of fractals and chaos theory to senescence. *JAMA, 267,*1806-1809.

Lipsitz, L.A.(2004). Physiological complexity, aging and the path to frailty. *Sci Aging Knowledge Environ 16,* 16.

Maddox, G., & Wiley, J. (1976). Scope, concepts and methods in the study of ageing. In R. Binstock and E. Shanas (eds.). *Aging and the Social Sciences.* New York Van Nostrend Reinhold Co.

Maddox, G.L. (1964). *The encyclopedia of aging.* New York: Springer Publishing Company, 254-5.

Maddox, G.L. (1987). *The encyclopedia of aging.* New York: Springer Publishing Company.

Marin, M. (2001). Successful ageing- dependent on cultural and social capital? Reflection from Finland. In I.Modi (ed.), *Ageing human development (pp. 145-159).* New Delhi: Rawat Publication.

Markus, D. (2001). Uniqueness of the Israeli over 70 years age group. In I.Modi (ed.), *Ageing human development (pp. 67-72).* New Delhi: Rawat Publication.

Masoro, E.J. & Austad, S.N. (2001). *Handbook of the biology of aging, 5th ed.* San Diego: Academic Press.

Miller, R.A.(2003). The biology of aging and longevity. In: Hazzard W.A, Blass J.P, Halter J.B, Ouslander J.G, and Tinetti M.E, (Eds). *Principles of geriatric medicine and gerontology,* (pp. 3-15). New York: McGraw-Hill.

Minosis, G. (1989). *History of old age.* Britain: Camelot Press.

Mitnitski, A.B., Song, X. & Rockwood, K.(2002). The estimation of relative fitness and frailty in community-dwelling older adults using self-report data. *J Gerontol Med Sci, 59A,*627-32.

Mullahy, P. (1970). *Psychoanalysis and interpersonal psychiatry.* New York: Science House.

Muller, S. F. (1994). Clinical presentation of depression in the elderly. *Gerontology, 40 Suppl 1,* 10-4.

Murphy, E. (1982). Social origins of depression in old age. *British J of Psychiatry, 150,* 801-530.

Myers, H. F. & Hwang, W. C. (2000). Cumulative psychosocial risks and resilience: A conceptual perspective on ethnic health disparities in late life. *US National Library of Medicine*: National Institute of Health.

Nair, G.R. (1982). Problems of the aged. *Social Welfare, 16(11),*13.

Nair, P. S. (1987). Effect of declining fertility on population of ageing. India Genius, 43, (34), 17-182.

Ozesmi, Y. (2006). Fuzzy cognitive maps of local people impacted by dam construction: their demands regarding resettlement. Department of environment engineering. *Ozesmi.org.*

Parsons, P.A.(2002*).* Life span: does the limit to survival depend upon metabolic deficiency under stress? *Biogerontology, 3,* 233-41.

Pendergast, D.R., Fisher, N.M., Calkins, E.(1993). Cardiovascular, neuromuscular and metabolic alterations with age leading to frailty. *Journal of Gerontology, 48, 61*-7.

Pope, A. & Tarlov, A. (1991). Disability in America: Toward a National agenda for prevention. *Institute of Medicine (U.S.) Committee on a National Agenda for the prevention of disabilities.* Washington DC: National Academy Press.

Prakash, I. J. (2001). On being old and female: some issues in quality in quality of life of elderly women in India. In I.Modi (ed.), *Ageing human development (pp. 333-341).* New Delhi: Rawat Publication.

Prasad, R. & Bould, S. (2001). Intergenerational households of Indian- Americans: The elders experience. In I.Modi (ed.), *Ageing human development (pp. 296-311).* New Delhi: Rawat Publication.

Ramachandran, V. (1981). Family Structure and mental illness in old age. *Indian J. of Psychiatry, 23(1),* 21-26.

Ramamurti, P.V. (1963). Problems of older people- Analysis of age trends. *Psychological studies, 15(2),* 128-130.

Ramamurti, P.V. (2001). Institutionalization of elderly and quality of life. In I.Modi (ed.), *Ageing human development (pp. 322-326).* New Delhi: Rawat Publication.

Ramamurti,P.V.(1970). Problems of older people- analysis of age trends. *Psychological studies, 15(2),* 128-130.

Ramasubrahmany, P. N. (1974). Old age does not come by calendar. *Social welfare, 21 (8),* 130.

Ramchandran, V. (1981). Family structure and mental illness in old age. *Indian Journal of Psychiatry,* 23 (1), 2-26.

Rani, P.M. (2001). Institutional care of the aged. In I. Modi (ed.), *Ageing human development (pp. 312-321).* New Delhi: Rawat Publication.

Rao (1989). India's grey sector: Psychiatry of old age in India.

Rao, A.v. (1979). Neuropsychiatry in Indian culture. *Canadian Journal of Psychiatry, 24(5),* 431-436.

Rao, V. (1981). India's grey sector. *Psychiatry of old age in India.* New Delhi: Rawat Publication

Ray,S.C. (1975). A medico-social study of aged persons. A Thesis for MD (P&SM).AIIMS, New Delhi.

Reid, I.R., Gallagher, D.J.& Bosworth, J.(1986). Prophylaxis against Vitamin D deficiency in the elderly by regular sunlight exposure. *Age Ageing 15,* 35-40.

Reisine, S. & Locker, D. (1995). Social, psychological and economic impacts of oral conditions and treatments. In LK Cohen (Ed.), *Disease prevention and oral health promotion (pp. 33-72).* Copenhagen: Munksgaard.

Rosenberg, I.H.(1993). Sarcopenia: origins and clinical relevance. *J Nutr, 127,*990-991.

Rosser, R. (1993). The history of health related quality of life in 10 1/2 paragraphs. *Journal of Royal Society for Medicine, 86(6),* 315-318.

Roubenoff, R. & Harris, T.B.(1997). Failure to thrive, sarcopenia and functional decline in the elderly. *Clin Geriatr, 13,*613-22.

Samuelsson, G. & Johnson,M.A. (2001). Centenarians in France, Georgia.Hungary and Sweden: social characteristics. In I. Modi (ed.*), Ageing human development (pp. 30-52).* New Delhi: Rawat Publication.

Seal, S.C. (1979). The problem of aged. *Indian J of public health, 23(2),* 61-62.

Seeman, T.E., Singer, B.H., Rowe, J.W., Horwitz, R.I. & McEwen, B.S.(1997). Price of adaptation: allostatic load and its health consequences. *Arch Intern Med, 157,*2259-68

Sen, A. (1993). Capacity and well-being. In Martha C. Nussbaum and Amartya Sen (eds.), *The quality of life (pp.30-53).* Britain: Oxford.

Shilling, L. E. (1984). *Perspectives on counselling theories.* NJ: Prentice-Hall.

Sherman, R. & Dinkmeyer, D. (1987). Systems of Adlerian therapy: *An Adlerian integration.* New York: Brunner/Mazel.

Sinha, D. & Jha, T. (1989). Invariance of mass and number among tribal and non tribal children: a study of the influence of age, sex, culture and habituation. *Journal of Personality & Clinical Studies, 5, (* 2), 105-114

Sinha, D. (1972). *The Mugal syndrome.* New Delhi: Tata McGraw Hill.

Sivakumar, D. (2001). Integration of strategies for the care of older persons: A cross- cultural study. In I.Modi (ed.), *Ageing human development (pp. 278-286)*. New Delhi: Rawat Publication.

Smith, D., (1973). *The geography of social well being in the United States*. New York.: Graw Hill,

Soltoft, F., Hammer, M. & Kragh, N. (2009). The association of body mass index and health-related quality of life in the general population: data from the 2003 Health Survey of England. *Quality of Life Research, 18,* 10, 1293-1299.

Spielberger, C. D., Sharma, S. & Singh, M. (1973). Development of Hindi edition of the state-trait anxiety inventory. *Indian Journal of Psychology,48,* 11-20.

Streib, G.F.(1983). The frail elderly: research dilemmas and research opportunities. *Gerontologist 23,* 40-4.

Suchindran, C. M.& Koo, H. P.(1997). Demography and public health. In Detels R, Holland W, McEwen J, Omenn GS, (eds). *Oxford Textbook of Public Health, 3rd ed, Vol. 2.* (pp. 830). New York: Oxford University Press.

Sullivan, H. S. (1953). *The interpersonal theory of psychiatry.* New York: W. W. Norton.

Sullivan, H. S. (1956). *Clinical studies in psychiatry.* New York: W. W. Norton.

Sweeney, F. J. (1975*). Adlerian counseling. Boston*: Houghton Mifflin Houghton Mifflin Company is a leading educational publisher in the United States. The company's headquarters is located in Boston's Back Bay. It publishes textbooks, instructional technology materials, assessments, reference works, and fiction and non-fiction for both young readers .

Sweeney, F. J. (1981). *Adlerian counselling.* Muncie: Accelerated Development.

Teyber, E. (2000). *Interpersonal process in psychotherapy (3rd ed.).* Pacific Grove, CA: Brooks Cole.

The World Health Organization (2008). *The global burden of disease: 2004 update* : Burden of disease in DALYs by cause, sex and income group in WHO regions, estimates for 2004. Geneva, Switzerland: WHO.

Tibbitts, C. (1960). Hand book of social gerontology: *Social aspects of ageing.* Chicago University of Chicago Press.

Tuckman, J., & Lorge, I. (1956). Perceptual stereotypes about life and adjustment. *Journal of Social Psychology, 43,* 239-245.

U.S. Census Bureau (1997)Release Date: June 9.

U.S. Census Bureau Population (2000). *Estimates by demographic characteristics.* Annual estimates of the population by selected age groups and sex for the United States: Population Division.

Vaarama, M., Pieper, R. & Sixsmith, A. (2004). Care-related quality of Life in old age: Concepts, models, and empirical findings. *The Psychiatrist, 28,* 411-414.

Van Staveren, W.A., de Graaf., C. & de Goot, L.C. (2002). Regulation of appetite in frail persons. *Clin Geriatr Med, 18,* 675-84.

Vaupel, J.W., Carey, J.R., Christensen, K., Johnson, T.E., Yashin, A.I.& Holm, N.V.(1998). Biodemographic trajectories of longevity. *Science, 280,* 855-60.

Verbrugge, L.M.(1991). Survival curves, prevalence rates and the dark matters therein. *J Aging Health, 3,*217-36.

Verheij, R. A. (1996). *Explaining urban-rural variations in health:* A review of interactions between individual and environment. *Social Science & Medicine,* 923-935.

Verkerk, & Karssing (2001). Health-related quality of life research and the capability approach of Amartya Sen. *Quality of Life Research, 10,* 49-55

Vita, A.J., Terry, R.B., Hubert, H.B. & Fries, J.F. (1998). Aging, health risks, and cumulative disability. *N Engl J Med, 338,* 1035-41.

Vollenwyder, N., Bickel, J. f., d'Epinay, C.L. & Maystre, C. (2001). Changing family networks and relationships in two Swiss Cohorts of elderly. In I.Modi (ed.), *Ageing human development (pp. 82-99).* New Delhi: Rawat Publication.

Vygotsky, L.S. (1978). *Mind and society: The development of higher mental processes.* Cambridge, MA: Harvard University Press.

Walker, A. & Mollenkop, H. (2007). *Quality of life in old age: Conceptual issues.* Netherland: Springerlink.

Wang, C. & Chan, C. (2009). Psychological adaptation status and health related quality of life among older Chinese adults with visual disorders. *Quality of life research, 18 (17),* 841-851.

Williams, F.M.,Wynne, H.,Woodhouse, K.W.& Rawlins, M.D.(1989). Plasma aspirin esterase: the influence of old age and frailty. *Age Ageing 18,* 39-42.

Wilson, I. B., & Cleary, P. D. (1995). Linking clinical variables with health-related quality of life: a conceptual model of patient outcomes. *J Am Med Ass,* 273,59-65.

Woodhouse, K.W., Wynn, H., Baillie, S., James, O.F. & Rawlins, M.D. (1988). Who are the frail elderly? *Q J Med, 68,*505-506.

Wynne, H.A., Cope, L.H., James, O. F.,Rawlins, M.D.& Woodhouse, K.W.(1989). The effects of age and frailty upon acetanilide clearance in man. *Age Ageing 18,* 39-42.

Zinta, R. L. & Tiwari, A. K. (2006). Relationship between self-efficacy, socio-economic status and performance in secondary students. *Research Journal Social Sciences, 14,* (1), 121-137.

Zinta, R. L., & Tiwari, A. K. (2007). Impact of guided mastery treatment on sjelf-efficacy among test-anxious students. *Journal of Psychological Researches, 51(*2), 91-101.

Zinta, R. L. (2008). Socio-economic status as a panacea for mental health: A note on displaced families of Bhakra Dam. *Nava Arthiki, 16, 2A, 49-74.*

Zinta, R. L., & Tiwari, A. K. (2008). Psychological analysis on project affected and likely to be affected families of SJVN Ltd. of Himachal Pradesh. *A project Reported submitted to the Integrated Institute of Himalayan Studies, HPU Shimla-*171005. (Recommended for publication).

Zinta, R. L., & Tiwari, A. K. (2008). Socio-economic and infrastructure facilities for sustainable rural development: A factorial analysis on Himachal Pradesh. A project Reported submitted to the *Integrated Institute of Himalayan Studies*, HPU Shimla-171005. (Recommended for publication).

Zinta, R. L. (2009). Bharat mein uch shiksha ki dasha aur disha. *Sanchar Bulletin, 1(1),* 104-106.

Zinta, R. L. (2009). Different facets of quality of life among the populace of mounting India in 21st century. *Akhil Bhartiya Shodh Seminar on Iksavin sadi mein Bharat: Dasha avam disha,* organized by University Grant Commission, New Delhi. Dayanand Vedic Snatkotar Mahavidyalaya, Urai, Jalon, U. P.

Zinta, R. L. (2010). Dimensions of quality of life in the tribal areas of Kinnaur: A caste-wise analysis. *A Research Project submitted to the Institute of Tribal Research and Studies, H. P. University, Summer Hill,* Shimla-171005.

Zinta, R. L., Manjul, M., & Kumar, D. (2010). Social status and emotional wellbeing among employed and unemployed tribal single women. *Nava Arthiki, 18(1A),* 1-27.

Zinta, R. L., Verma, L. K., Thakur, R. (2010). Inquesting psychological wellbeing among the squatters of Shimla town. *Nava Arthiki, 18(1A),* 61-91.

Zinta, R. L. (2010). *Psychology Manual.* New Delhi: H. G. Publications.

Zinta, R. L. (2010). Performance among high and low self-efficacious students. In P, Ramalingam (Ed.), Recent studies in school psychology. Delhi: Authorspress.